ON TIME

ON TIME

A History of Western Timekeeping

KEN MONDSCHEIN

Foreword by Neal Stephenson

JOHNS HOPKINS UNIVERSITY PRESS | *Baltimore*

© 2020 Johns Hopkins University Press
All rights reserved. Published 2020
Printed in the United States of America on acid-free paper
9   8   7   6   5   4   3   2

Johns Hopkins University Press
2715 North Charles Street
Baltimore, Maryland 21218-4363
www.press.jhu.edu

Library of Congress Cataloging-in-Publication Data

Names: Mondschein, Ken, author.
Title: On time : a history of western timekeeping / Ken Mondschein ;
    foreword by Neal Stephenson.
Description: Baltimore : Johns Hopkins University Press, 2020. |
    Includes bibliographical references and index.
Identifiers: LCCN 2019040487 | ISBN 9781421438276 (paperback) |
    ISBN 9781421438283 (ebook)
Subjects: LCSH: Clocks and watches—History. | Time measurements—History.
Classification: LCC TS542 .M654 2020 | DDC 681.1/1309—dc23
LC record available at https://lccn.loc.gov/2019040487

A catalog record for this book is available from the British Library.

*Special discounts are available for bulk purchases of this book. For more
information, please contact Special Sales at specialsales@jh.edu.*

Johns Hopkins University Press uses environmentally friendly book materials,
including recycled text paper that is composed of at least 30 percent post-consumer
waste, whenever possible.

To my father, Michael Mondschein
I wish our time together had been longer

CONTENTS

List of Illustrations   ix
Foreword, by Neal Stephenson   xi
Acknowledgments   xiii

1   Scholars and Spheres   1
2   Cities and Clocks   39
3   Savants and Springs   75
4   Navigators and Regulators   110
5   Rationalization and Relativity   147

Appendix. Chapter Exercises   179
Glossary   199
Notes   209
Suggested Further Reading   223
Index   235

1.1. The Ptolemaic model of the universe 18
2.1. Parts of an astrolabe 45
2.2 Vitruvian water clock 48
2.3. Astrarium 62
2.4. Virge and foliot 72
2.5. Su Song's clock 59
3.1. Santorio's pulsilogium 82
3.2. Huygens's clock 90
4.1. Analemma 118
4.2. The equation of time 119
4.3. The II4 124
5.1. Time dilation 153
5.2. Atomic clock diagram 171
5.3. 1950s atomic clock 171
A1.1–1.6. Analemma 181
A2.1–2.4. Astrolabe parts 187

Time seems obvious to most people and is fundamental to science and technology, but the more one thinks about it, the less one really understands. Is it an intrinsic property of the universe or an epiphenomenon of the way we think? Can we conceive of a physics that does away with time altogether, or should we content ourselves with a merely elastic account of time, as predicted by the theory of relativity and verified by many experiments? These questions have intrigued and bedeviled some of humanity's finest minds. Consequently, the topic can seem forbidding.

When trying to follow a scientific account of the universe, rather than tackling mathematical abstractions head-on, historical narrative can be an illuminating way to understand what problems scientists were actually trying to solve and how they ended up pursuing the lines of inquiry that led them into those abstractions. In this volume, Ken Mondschein ably uses that strategy to tell the story, and trace the development, of time-related thinking from ancient times onward in a way suitable for both undergraduate historians of science and for general readers. While enjoying a series of well-told anecdotes about historical figures trying to solve practical problems, the reader is introduced to the fundamental philosophical questions that arise around time—and given a short course in basic astronomy to boot. We learn how the idea of time became fundamental to classical scientific thought and how the technologies needed to craft useful clocks drove other developments in materials science and mechanical engineering. The story is concluded with a series of practical exercises useful for helping the reader develop some intuition about all of these concepts. Mondschein ends with an overview of recent and current developments, demonstrating that the field of time and timekeeping is as active and as central to modern science and technology as it was to ancient and medieval natural philosophers.

Neal Stephenson

ACKNOWLEDGMENTS

First, of course, I have to thank Tom Glick, my mentor at Boston University, who first awakened me to the connections between the Islamic world and Western science. The timekeeping project took shape in Dan Smail's graduate seminar, and his way of thinking about space lent a lot to my thinking about time. Richard Gyug was the best graduate advisor I could have asked for. Bert Hall's guidance and advice in the early phases of peer review was invaluable. Thanks to Neal Stephenson for contributing the foreword and Steven Walton for additional proofreading.

Libby Craun and Sarah Dean kept me fed and sane. Finally, much appreciation to Greg Britton, Matthew R. McAdam, and William Krause at Johns Hopkins University Press and copyeditor Carrie Watterson. Without their contributions, the book would consist of blank pages.

ON TIME

ONE

Scholars and Spheres

Far away across the field,
The tolling of the iron bell,
Calls the faithful to their knees,
To hear the softly spoken magic spells.
—Pink Floyd, "Time," *The Dark Side of the Moon*, 1973

CAPTAIN DIMITRIOS KONDOS and his crew could hardly believe their shipmate Elias Stadiatis's claim: An ocean floor full of nude corpses lay rotting 60 meters, or almost 200 feet, below the surface of the Mediterranean. The men were sponge divers, earning their living by an age-old, but perilous, craft. Though they used what was, for the year 1900, rather sophisticated equipment—copper helmets and canvas diving suits—decompression sickness and nitrogen narcosis, which can cause delirium, were ever-present dangers. Added to this were the timeless hazards of navigating the Mediterranean—most especially, storms. The need to wait out just such a tempest was the reason Kondos and his crew were anchored here off the island of Antikythera, halfway between Crete and the Peloponnesus and far from their usual diving grounds.

Even though the effects of the deep dive seemed to be the most likely explanation for Stadiatis's fantastic claims, Captain Kondos himself decided to go down. When he reemerged from the deep, he was holding the bronze arm of an ancient statue. He and his men had stumbled upon one of the richest underwater archaeological finds ever discovered: a ship on its way from Rhodes to Rome with a cargo of fine art for the new masters of the Mediterranean world had gone down sometime in the first century BCE, scattering Greek statues, coins, jewelry, and pottery on the sea floor.

The greatest treasure recovered from the Antikythera wreck, however, was not a statue or work of fine art but a seemingly unremarkable piece of corroded bronze. While at first the artifact was overlooked in favor of the more obviously valuable sculptures and coins, in 1902, an archaeologist named Valerios Stais noticed that the fragment had a large wheel embedded in it. Stais argued that the piece was some sort of clock, but his interpretation was dismissed. A clock would be too advanced, the thinking went, for the time period from which the wreck dated. After all, scholars were in agreement that sophisticated timekeeping instruments had been invented only in the late Middle Ages.

The enigmatic artifact remained little known and little studied until the British physicist and historian of science Derek John de Solla Price became interested in it in the 1950s. In the 1970s, together with the Greek nuclear physicist Charalampos Karakalos, de Solla Price took X-ray and gamma-ray images of the mysterious piece of bronze. The results were astounding: the fragment was full of hand-cut gears. By analyzing the size and the ratio of these gears—that is, how much each moved in comparison to the others—de Solla Price and other researchers were able to hypothesize the purpose of this ancient machine, dubbed the Antikythera Mechanism.[1]

The Antikythera Mechanism is an immensely sophisticated analog computer, composed of numerous precisely designed gears. (An analog computer is one that uses a continual physical phenomenon, be it mechanical, hydraulic, or electric, to record data; unlike a digital computer, it is not programmable.) When the mechanism's interlocking gears, which are estimated to have numbered from 30 to 50 when it was complete, were cranked, they would move dials to represent the position of the planets, calculate calendars, predict solar eclipses, and determine the ancient Olympiads (the four-year ritual period between the ancient Olympic games).[2]

The Antikythera Mechanism is, in other words, an early example of two goals toward which human beings have applied a great deal of effort, energy, and ingenuity: the precise knowledge of time and the simplification of the complex process of obtaining that knowledge. Mechanically modeling the natural cycles of the sun, the stars, and the moon—the foundational measurements of timekeeping—not only told its user facts about the heavens but predicted the future. Plus, a mechanical device would be far easier to consult than following the different calendars in use in the ancient world or than making complicated calculations and observations of the heavens. This is the very stuff of technology—the application of knowledge and inventiveness to save labor and gain power over the natural world.

Just as astronomy is the primary science that led to the idea that the natural world functions by knowable mathematical principles, timekeeping is the primary *technology* on which all others depend. While the year, the days, and the night are obvious natural divisions of time, to create a calendar reconciling the lunar and solar year, or to divide the day into 24 equal parts—or to subdivide those units into minutes and seconds—is less obvious and requires some careful thought. Even more difficult is to create a machine that automates this work. Yet clocks were some of the first sophisticated machines to be invented. Eventually, the machine itself came to stand for the thing it modeled. The clock became the emblem of time, replacing the natural processes it modeled.

Along with this came a shift in mentality. Today, the mechanical reckoning of artificial time—that is to say, the Western idea of time reckoning—seems obvious. We have been governed by mechanical clocks since birth, the hour of our entrance into his world recorded on official documents. From our childhoods, we have lived according to the clock, dropped off and picked up at school, at friends' houses, and at after-school activities at designated times. All college students know the feeling of watching the minute hand creep toward the moment the professor will release a long-suffering class, and, in the working world, the time clock rules all. Of course, these hours are totally artificial—the same for everyone in the time zone, no matter what the position of the sun might be in a certain location. The actual local time in Kalamazoo, Michigan, for instance, is more than an hour later than in Bangor, Maine, even though both keep "eastern time."

Science is dependent on the accurate measurement of time. On the largest scale, we know how far *Voyager I* and *II* have traveled from

Earth by how long it takes their radio signals to reach us. On the smallest, to observe subatomic particles such as the Higgs boson, scientists must be able to divide not only matter but also seconds into tiny fragments. Time and space are inextricably related: the time signal transmitted by global-positioning satellites is triangulated by handheld devices and used to compute the location of everything from jet airliners to lost hikers.

All of this depends on universally recognized and agreed-upon time measurement. At first, these were visual observations of obvious phenomena such as the sun going down or a certain constellation appearing on the horizon. People quickly developed tools to sight heavenly bodies or measure other observable phenomena, such as the shadow of a sundial. As human knowledge grew, increasingly complicated devices such as astrolabes and mechanical clocks came into use. The quest for increased precision and accuracy meant that innovations such as Galileo's pendulum clock would yield to Robert Hooke's springs and, finally, the atomic clock. (*Precision* is the ability to measure fine divisions of time, and *accuracy* is the ability to do so consistently.) The more reliable, precise, and accurate these mechanisms became, the more they came to be seen as the primary indicators of time, rather than as the observations of the natural world such measurement is based on. In the nineteenth and early twentieth centuries, mechanical time came to be primary, with a "standard time" for a nation replacing local solar time. Rather than clocks being calibrated to the natural world, the natural world itself came to be measured by mechanical devices—so much so that today the atomic clock has assumed primacy over the celestial phenomena that were once the metric and emblem of time itself. Thus, the way in which people *thought* about time changed alongside how they measured and used time. The tale of modern scientific timekeeping is a tale of increasing precision—and placing that precision in the hands (and on the wrists) of everyday people had inevitable consequences for how we think about time and how we run our society.[3]

This is a book about the development of Western timekeeping technology and its interplay with the histories of ideas and culture. Although I will touch upon the innovations of other cultures (Western timekeeping especially owes a profound debt to medieval Arabic science) and offer some comparative histories in sidebars, the systems of China, Mesoamerica, and Africa are by and large outside the scope of this book and this series. Besides, since I am neither an Arabicist nor a Sinologist, I cannot do these subjects justice—and in any case, including all this information would make this book unmanageably large and sprawling. Still, even while keeping our focus, I want to offer a word of caution about imagining "progress" as inevitable or the Western way of doing things as superior. Despite this caveat, the story of how the world came to be governed by European ideas of time is informative, because it is also the story of imperialism—an imperialism not of "guns, germs, and steel," as Jared Diamond put it,[4] but of ideas.

Likewise, for many previous writers on the history of science and technology, it was innovation, spurred on by farseeing "geniuses," that drove social change from the supposed depths of medieval ignorance to the alleged glories of modernity. But we can just as easily see things from the opposite perspective: a perceived need precedes innovation. Email, for instance, has transformed the way we communicate, but Ray Tomlinson "invented" the medium in 1971 as an extension of a local messaging program known as SNDMSG that had already proved useful, and its subsequent widespread adoption was but a logical continuation of the sorts of demand that had made teletype and telegraph machines indispensable to newsrooms and stock exchanges.[5] Similarly, the utility of the cellular phone was presaged by the World War II–era radio telephone and the bulky and expensive (but prestigious) car phone; however, it was not until the late twentieth century that infrastructure and technology caught up with consumer demand.

I would argue that it was the fact that people in Europe had the *desire* for precise and accurate but easy-to-use timekeeping and time-signaling technology that led to technological and social change. The peoples of Asia, the Americas, the Pacific, the Middle East, and Africa all had their own motivations and developed their own methods of timekeeping that worked within the contexts of their societies. However, imperialism and the perceived need to "modernize"—to learn Western scientific and industrial methods and run a nation's military and economy according to

What Is a Clock?

Considering that this is a book about timekeeping, it might be worthwhile to define what, exactly, a clock is. Simply put, it a machine that takes a continuous physical phenomenon that happens at a known rate—be it gravity pulling on a weight or causing water to flow from a reservoir into a receptacle, the unwinding of a spring, or electronic impulses—and makes how much of that phenomenon has occurred readable to humans—whether by a measure of liquid, hands on a clockface, or a digital display—so that an observer can know how much time has passed. In this, it differs from tools such as sundials and astrolabes, which are observational devices used to show the passage of time as indicated by natural phenomena such as the movement of sun or stars. A sundial, for instance, allows us to observe and measure the movement of the sun through the sky, whereas a clock allows quantification of an independent, human-controlled phenomenon. A clock, in other words, is entirely artificial, whereas the sundial or astrolabe depends on natural phenomena to make it "work."

their precepts—spread the European way of doing things. Although in hindsight the techniques developed by the West may seem inevitable, this is only an illusion wrought by the legacies of imperialism.

What Is Time, and Why Do We Measure It?

The early Christian theologian Augustine of Hippo (354–430) asks in his *Confessions*, "What then is time? If no one asks of me, I know; if I wish to explain to him who asks, I know not."[6] A modern person hearing this question might simply point to a clock as a way of explaining "what is time," but then what does a clock measure? When we look at a clock, we aren't really defining time—we're observing how the clock has quantified whatever phenomenon it is measuring (for instance, how much the hands of a grandfather clock have moved because of the weights being pulled down by gravity). There's a deep truth here: the *only* way we know time passes is by observing change. If you doubt this, think of falling deeply asleep or going under anesthesia for surgery. The next thing you know, you're awake—the time in between is "lost." Similarly, people placed in a sensory deprivation tank lose the ability to sense time.

To use an old metaphor, time is a fog-shrouded river where we can't see either bank. Without a stable reference point, we can perceive our own motion only by watching other things pass us by—floating branches carried downstream by the current, rocks slipping by, other boats rowing upstream. In other words, we can tell time is passing only by watching changes in the world, be they the rising and the setting of the sun, the passage of the seasons, or the movement of our own thoughts. The idea of time progressing at a constant rate is no more than a convention—a convention contrary to most human experience but nonetheless highly convenient for regulating activities such as work, factory production, and trade. For this happen, a mental shift had to occur, from people considering time as the comparison of relative durations to time as a measure of some absolute metric that exists independent of any human experience. To do this required increased precision and accuracy. In one sense, then, we can see the history of timekeeping as a march toward more perfectly describing our drift in time's current.

But the growth of precision and accuracy is not the only theme in the history of timekeeping; equally important are *simplification* and *ease of use*. The cycles of the natural world are complicated and difficult to understand, while a clock (or a calendar) is intended to be simple. Using the Antikythera Mechanism, for instance, would have been much easier than a process of meticulous astronomical observations and calculations.

Other questions we will address include *how* and *why*: How and why did this idea of exact, yet abstract, time reckoning arise? What brought about this innovation? Why were people so interested in it, and how did it, in turn, come to shape science and thought? How did the idea of living by artificial measurements of time come to dominate society in Europe, and how was this similar to and different from time regimes in other regions? And what were the scientific, technological, social, cultural, and even philosophical effects of this mental leap?[7]

These are the questions this book will answer. In this first chapter, I look at astronomy, which is the physical and theoretical basis of all timekeeping and calendar systems, both Western and non-Western. First, the chapter examines the astronomical beginnings of ancient systems of timekeeping. Then it looks at some surprising facts about how we reckon time from the sun, moon, and stars. Finally, it explores astronomical timekeeping in the ancient world and the inheritance the scholars of the Middle Ages received from their classical forebears.

In the second chapter, I continue our look at medieval timekeeping, leading up to the invention of the mechanical clock in the late Middle Ages as the mirror of the heavens. I'll consider how this was reflected in medieval philosophy, which saw time as the comparison of relative durations—specifically, the movement of the sun, moon, stars, and other astronomical bodies. The third chapter examines the inventions of the seventeenth and eighteenth century, how they were based in medieval thought (though they quickly outgrew its strictures), and how these precise timekeeping devices accord with Isaac Newton's idea of "absolute time"—that is, abstracted time divorced from the measurement of observable phenomena. Rather, like the hands of a clock, time marches on independently of seemingly anything else. The fourth chapter begins with the invention of the most precise timekeeping device yet, the longitude chronometer, which seemed the very stuff of Newtonian absolute time made tangible and created the idea of an abstract "standard" time. Then, in the second part of chapter 4, I look at how the internalization and standardization of this abstracted time, detached as it was from observable phenomena, came to regulate industrial society, and I explore how this "standard time" spread throughout the whole world. Finally, the fifth chapter begins with Einstein's disruption of the Newtonian paradigm and then looks at the modern technology that has made possible precise and accurate timekeeping—so much so that timekeeping has been utterly sundered from its original astronomical roots. Finally, I consider how the urge to standardize and automate society according to machines almost led to disaster with the Y2K problem of the turn of the millennium. But to understand how one paradigm shifted to another, we need to know the entire story. Let us therefore begin at the beginning, with the astronomical basis of timekeeping.

Origins of Systems of Timekeeping

We humans share with birds, plants, and even microbes an "internal clock" that regulates biological activities such as sleeping, waking, and reproduction. Together with this, we have an inherent conception of *timing* that derives from a sense of rhythm and probably evolved in our ancestral environment because of the advantages it confers. A highly developed sense of rhythm helps a tribe function as a single unit in war and in hunting. Work songs, hunting songs, and ritual chant all enable a group of individuals to labor in harmony, melding themselves into a combined organism that is more than the sum of its parts.[8]

We humans are also unique in our abilities to predict the future based on observations of the external world and to use technology to extend our own senses. Even hunter-gatherer societies have these skills. For instance, the British historian of science G. J. Whitrow describes how Australian aborigines would agree to perform some action when the sun struck a stone placed in the fork of a tree.[9] This is the second sort of conception of time: measurement, whether of duration (how long something lasts) or timekeeping in the proper sense (the counting of arbitrary intervals). Both of these mental structures require some sort of abstract division of time—in other words, enumeration.

The use of number in human timekeeping mirrors the periodicity of the natural world: a day is a cycle of light and darkness; a month is a full cycle of the moon, from new to full and back to new; a year is a full turn of the seasons, with their associated changes in the position of the stars, the weather, and the length of day. All timekeeping in the modern, scientific sense—which is to say all notions of regular time—originates from the observation of the heavens. After all, the stars and other celestial bodies are the most predictable and regular moving objects in the natural world. While the origins of artificial, human-created divisions of time are less obvious, applying mathematical analysis to these natural movements would seem to be the next logical step. For instance, the idea of dividing both day and night into 12 parts is so ancient that we cannot identify a definite starting point. The Egyptians, the Babylonians, the Chinese, and the civilizations of the ancient Indus

Valley all used duodecimal (that is, base-12) counting systems. The use of the number 12 seems counterintuitive because today we use base-10 in everyday life, but it probably comes from the number of finger bones in the human hand. It also conveniently mirrors the approximate number of lunar cycles in a year. Despite its arbitrariness, this system has a deep hold on us. The French Revolution introduced not only metric weights and measure but also metric time. The meter, liter, and kilogram are still around, but decimal time and the 10-month calendar are mere curiosities, at least in the West.

There also exists a third conception of time: that of myth, legend, and theology, whether the biblical times of creation in Genesis and judgment in Revelation, the era of the Greek gods, the Hindu Vedic ages, or medieval Christian theorizations of time, eternity, and sempiternity. (A sempiternal thing has a beginning but no end; for instance, while God is eternal, angels, which are immortal created beings, have a beginning but no end.) Although such concerns are not as relevant to the measurement of time in the sense of concrete observations of physical phenomena, as we'll see, they did open up important mental vistas that eventually led to the theoretical world of physics. By creating "imaginary time," thinkers created the possibility of abstract theorization, which, in turn, led to new ways to create models that would predict how real phenomena occur in the real world.[10]

Theological time is also where we get our dating of BC and AD, though the secularized modern usage, as in this book, is "Common Era," or CE, and "Before the Common Era," or BCE. AD, for *anno Domini* (year of the Lord) was first introduced by the widely influential English monk Bede (c. 672–735 CE) in his *Ecclesiastical History of the English People* (*Historiam ecclesiasticam gentis Anglorum*), dating from the conception of Jesus as calculated by the sixth-century CE monk Dionysius Exiguus.

Finally, theological time enters into another aspect of timekeeping: the awe, wonder, and sense of smallness that human beings feel when confronted with the immensity of the night sky all lend themselves easily to religious feeling. When we combine this with the powers of social coordination and prediction inherent in timekeeping, we can see why the interpretation of the movements of celestial objects and, thus, the reckoning of time have historically been connected with religious practice in human cultures. To predict the future is to control the actions of others. In many cultures, astronomy and timekeeping have historically been privileged knowledge, restricted to the few. Even in the modern world, we have authorities to tell us what time it is—even if the authorities take their ultimate cue from a machine.

The First Astronomy and Calendars

Calendars probably predate civilization itself. In fact, if you think about it, we still use calendars for the same purpose that prehistoric people did—social coordination. This is useful not just for religious reasons—When is Christmas? When is Easter? When is Ramadan? When is Yom Kippur?—but also more everyday ones—When does the school semester begin? When is summer vacation? When is the movie I've been looking forward to going to be released? These are useful dates to know, but in a hunter-gatherer society, knowing the cycle of the year would have been critical to survival itself: When will the animals migrate? When will the green plants come back? When will the river flood?

People have always used observations of the natural world for these purposes. Archaeologists have found tally sticks used to count the lunar cycle that date to before the invention of agriculture, and the configuration of the stars in the constellation we know as Orion appears in an image carved into mammoth ivory between 32,000 to 38,000 years ago and found in Germany's Ach Valley in 1979.[11] This implies that early people watched the heavens to note the change of seasons, because different constellations are visible at different times of the year owing to our changing perspective as the earth goes around the sun. Orion, for instance, is not visible in the Northern Hemisphere from May to July. Of course, the stars are still there, but they rise during the day and so are outshone by the sun. (If there is a solar eclipse during this time, though, we can see Orion.)

The Neolithic revolution, that is, the invention of agriculture, would have only accelerated this tendency: as societies developed farming, knowing the cycle of the year and planning labor accordingly was critical to survival. The ancient Egyptians, for instance, organized their year around the annual summer flooding of the Nile, which was key to their system of agriculture. Religious festivals also inevitably follow the cycle of the year; for instance, holidays marking the winter and summer solstices (the shortest and longest days of the year) are nearly universal. Determining the solstices is in fact the purpose of the oldest known surviving application of technology for marking time—the creation of permanent markers.

Prehistoric monuments such as Stonehenge in Wiltshire, England; Zorats Karer in Armenia; and Chanquillo in Peru are often held up as examples of ancient timekeeping devices.[12] There is no agreement as to what, if anything, these monuments were supposed to mark.[13] In all likelihood, it wasn't the same thing, since they were associated with rituals that had meanings particular to the widely disparate cultures that produced them. However, other ancient sites, such as the several hundred circular enclosures constructed in central Europe approximately 6,000 years ago (4900–4600 BCE), were definitely aligned with calendar observations. For instance, at the circular enclosure archaeologists found at Goseck in the central German state of Saxony-Anhalt, openings in the palisade were aligned with the directions the sun rose and set at the winter solstice. These were also probably associated with religious observances, since human and animal bones, including a headless human skeleton, were also excavated at the site.[14] Even among far-flung early human civilizations, forms of timekeeping were definitely used—even if we can't always figure out how they worked.

Written records came somewhat later, but they had the advantage that detailed observations could be conducted over a long time, even generations, thus showing long cycles of natural phenomena and enabling more precision. As early as the 1600s BCE, the ancient Babylonians, who ruled a kingdom in Mesopotamia (the land between the Tigris and Euphrates Rivers in modern Iraq), were recording observations of Venus.[15] The Egyptians took similar measurements: the ceiling of the Senenmut Tomb from Egypt's 18th dynasty (c. 1473 BCE) depicts the constellations by which the three four-month-long Egyptian seasons of Akhet (when the Nile flooded), Peret (planting), and

Shomu (harvesting) were reckoned.[16] The Egyptians also had, by 1500 BCE, sundials for reckoning time during the day (though obelisks, which could be used for similar purposes, date back to 3500 BCE), a *merkhet* (a device consisting of a hanging weight) for sighting stars to tell time at night, and water clocks to measure intervals at any time.[17] Like the Mesopotamians, the Egyptians divided the day into 12 parts.

The accumulation of observations allowed ancient astronomers to note the regularity of certain phenomena. For instance, the cuneiform tablet known as MUL.APIN (from the first words on the tablet, which was the name for the Plow Star in Akkadian, the language of the Babylonians) dates from the 700s BCE. MUL.APIN is a star chart that coordinated the visibility of certain stars with the twelve-month calendar. As an example, on the first of the spring month Nisannu, the Hired Man (the constellation we call Ares) rises; on the 10th of Simanu, the True Shepherd of the sky god Anu (that is, Orion) becomes visible. It also tells how many days pass before the rising of certain constellations— for instance, 30 days from the Scales (Libra) to the She-Goat (which corresponds to none of our constellations but contains parts of Lyra and Hercules).

Why does MUL.APIN give *both* the dates and the intervals between constellations rising? The reason is that the Babylonians hadn't yet accumulated a sufficient number of observations to allow for the creation of definite mathematical rules for a calendar. Thus, they couldn't have a real division between the ideas of "measuring duration" and "timekeeping." Instead, they had a number of "rules of thumb" to use if the natural signs were too out of step with the calendar. For instance, if the Stars (the star cluster we call the Pleiades) rose on the first of the month Ajjaru, it was a normal year; if on the first of Simanu, they would add a 13th "leap month" to bring the calendar back into order.

MUL.APIN also tells which zodiac signs set as others rise. For example, the Scorpion (Scorpio) sets as the Stars rise. This way, if the sky were cloudy in the east, one could tell the time by the sign setting in the west. Similarly, knowing what signs will appear when is convenient if you are tired and don't want to stay up all night watching the stars. Thus, we see one of the motivations for mathematical tools and, eventually, mechanical means of timekeeping to model the heavens: not only the desire for greater precision but also for convenience's sake.[18]

The Babylonian astronomers' observations not only guided the agricultural calendar but were believed to foretell the future. For instance, if the sun looked "sprinkled with blood" when it rose on the first of Nisannu, then there might be a famine or the king might die.[19] This seemed quite logical; in fact, Europeans believed the heavens could affect the weather well into the seventeenth century CE. After all, it stood to reason that phenomena that happened in the sky were all related. Likewise, astrology and astronomy were not considered separate spheres of knowledge—it was taken for granted that the stars could affect what happens on Earth. Astronomers thus had an important place in Babylonian society: theirs was the power of prophecy. We must always remember that timekeeping originated with an attempt to better understand an unpredictable universe and was connected to the divine order of the world.

Ancient Greek Timekeeping

Compared to Mesopotamia or China, Greece in the eighth century BCE was a comparative backwater just rediscovering writing as it emerged from a protracted dark age. In the long narrative poem *Works and Days*, which was probably handed down orally for many generations before it was written down by the poet Hesiod around 700 BCE, we have a list of traditional astronomical observations, as well as other natural signs, by which the ancient Greeks knew the labors of the year. Hesiod tells us that "when the Pleiades, daughters of Atlas, are rising, begin your harvest, and your plowing when they are going to set," and that, "when Zeus has finished sixty wintry days after the solstice, then the star Arcturus leaves the holy stream of Ocean [that is, rises above the sea] and first rises brilliant at dusk. After him . . . the swallow appears to men when spring is just beginning. Before she comes, prune the vines."[20] Like the Babylonians, the Greeks made much of the meanings of heavenly signs, but they lack the compiled detailed observations of Mesopotamian astronomy. We get the image of an impoverished, conservative society in which the stars and other natural signs were mainly heeded to avoid offending the gods and to know how best to eke out a scant living from the poor soil. Yet, by the first century BCE, the Greeks had absorbed and built upon the learning of the Mesopotamians and Persians, whom they had conquered under Alexander the Great. Greek timekeeping technology became the marvel of the world, culminating in devices such as the Antikythera Mechanism. Even more importantly, the Greeks compiled methods of using mathematics and geometry to create a theoretical model that could predict the movement of heavenly objects and thus tell time with both precision and relative simplicity. We must not imagine that the Greeks were the only ones to do this—Chinese and Indian astronomers had achieved similar accomplishments. However, it was through Greek (and to a lesser extent, Roman) sources, transmitted from the Arabic-speaking world, that the idea of mathematical timekeeping passed to the medieval West and formed the foundation for later developments.

Our greatest source for Greek achievements in astronomy is Claudius Ptolemy. Ptolemy was born in Egypt around 90 CE and died in about 168 CE, some two or three centuries after the Antikythera Mechanism was constructed. At the time he lived, Egypt was a territory of the Roman Empire, a crossroads of ancient Mesopotamian astronomical learning, Greek scholarship, and Latin power. Like other educated people in the eastern Mediterranean, Ptolemy wrote in Greek—thus the title of his great work on the heavens came to be known, *Iē megalē syntaxis*, or *The Great Treatise*. Judging from the astronomical observations he included, Ptolemy finished this work sometime between 147 and 161 CE. Ptolemy also wrote the *Tetrabiblios*, a work on astrology, and the *Geography*, which was the standard reference work for what people knew about the world, such as the division of the three classical continents (Europe, Asia, and Africa), before the age of exploration in the sixteenth century CE.[21]

Because of the international scope of the Roman Empire, Ptolemy's astronomical textbook was diffused all through the Mediterranean world. In fact, it was so successful that it displaced almost all earlier works—meaning that much of what we know about ancient astronomy (and, thus, timekeeping) comes from what Ptolemy told us about his predecessors. Even though it was almost lost when literacy rates, and interest in the natural world, plunged when the Western Roman Empire dissolved in the fifth and sixth centuries, it was both preserved in the Greek East and translated into Arabic. It was from the Arabic translation that this important work was later reintroduced to medieval Europe. In fact, the Arabic transliteration of the title of *The Great Treatise*, *al-majistī*, gives us the name by which Ptolemy's book has come to known in the West—the *Almagest*. The *Almagest* remained the standard astronomical technical treatise until Copernicus developed his heliocentric model of the universe in the sixteenth century CE.[22] (Copernicus published in 1543, but it would take more than a century and a half for all European astronomers to accept his ideas.)

In Ptolemy's model of the universe, the earth is positioned in the middle of an immense, invisible sphere, like a marble floating inside a snow globe. On the outer surface of this sphere are mounted the stars visible in the night sky. The planets, the sun, and the moon are all mounted on their own spheres, which rotate around the earth at their center. This is called the *Ptolemaic* view of the universe, and, though it was named after Ptolemy, it was also famously described by the Greek philosopher and natural scientist Aristotle (384–322 BCE), whose work formed the basis of physical science up until the time of Galileo. Even though we know today that the earth and other planets go around the sun, for centuries Ptolemy gave people a model that worked well enough for their purposes, and modern astronomers still use an imagined Ptolemaic universe to describe the positions of celestial bodies as seen from Earth.

Because the earth rotates on its axis, the stars (or, rather, the imaginary sphere they are mounted on) will seem to revolve in a complete circle once every 23 hours, 56 minutes, and 4 seconds (counting, of course, with our modern hours, minutes, and seconds). The unit of time marked by the revolution of the stars is known as the *sidereal day*. It may seem odd that the period of the earth's daily rotation is 3 minutes and 56 seconds shorter than a full day, but the length of the *mean solar day*—the unit of time we refer to when we say a "day" in ordinary language—is extended because the earth is constantly traveling around the sun. Because the earth traverses about 1/365 of its orbit every day, it has to rotate a little extra until the sun "catches up" to where it "ought to be" in the sky relative to an observer standing on the earth's surface. Of course, clocks and timekeeping had to be fairly well developed before the difference in times could be observed, and, for most of history, the sidereal day *was* the "day," and 1/24 of this was the "hour." How modern hours came to be reckoned, and how the technology for doing so was developed, is a complicated process that I will discuss throughout the rest of this book. In the context of ancient and medieval timekeeping, however, we can define a day as one revolution of the earth—which is to say, the sphere of the stars.

The revolution of the "fixed stars" was the most regular heavenly movement, because the period it takes for a given star to perform one revolution and come back to the same altitude over the horizon will not change appreciably during a human lifetime. In other words, the sidereal day is seemingly fixed, regular, and unchanging. This is why thinkers in the premodern world considered it the most basic movement and thus the fundamental unit of time and the basis for "true" horology. (*Horology* is the science of telling time.) It was not until much later that people began to use abstracted hours based on more sophisticated observations.

Of course, there are thousands upon thousands of stars visible in the sky, so which ones are we to watch? Ptolemy followed his Mesopotamian predecessors by organizing the brightest stars in the sky into mnemonic star patterns, called the zodiac (from the Greek *zōdiakos kyklos*, or "circle of animals"). In fact, the 12 constellations Ptolemy mentioned are the same ones we use today. However, because of a phenomenon called *precession* caused by changes in the earth's axis and orbital path, the constellations are not actually in their original zodiac signs today.

Not only do the 12 constellations match the twelve hours in the day, the 12 bones in the human hand, and the approximately 12 lunar cycles in a year, but, because measuring angles in the sky is so fundamental to timekeeping, we have also used base-12 math in both timekeeping and geometry since the time of the Babylonians. For instance, $12 \times 5 = 60$, the number of minutes in an hour, and $12 \times 30 = 360$, which is both the number of degrees in a circle and the approximate number of days in a

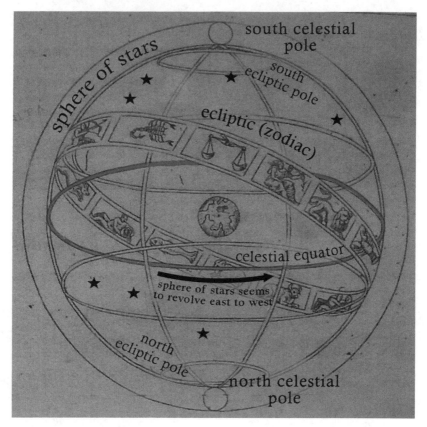

Figure 1.1. The Ptolemaic model of the universe, adapted from an illustration in Camillo Agrippa's 1553 *Treatise on the Science of Arms*. Note that south is at the *top*, which was not uncommon in Renaissance astronomical diagrams. The fixed stars seem to rotate around the earth's axis once per sidereal day along paths parallel to the celestial equator. The zodiac signs are located on the ecliptic, the path of the sun's apparent motion against the backdrop of stars. Either the ecliptic or the celestial equator can be used as the basis for a coordinate system to locate astronomical objects. Images from Agrippa courtesy of Malcolm Fare

year. There are two twelve-hour halves in the twenty-four-hour day; because 360 ÷ 24 is 15, the celestial sphere will rotate 15 degrees in an hour as the earth turns on its axis, which is how we can tell the hour from the stars (as I'll discuss below). This math also means we can specify a location on the earth or the heavenly sphere by giving latitude and longitude coordinates in degrees and minutes, remembering that 15 degrees are equal to an hour; minutes, represented with a prime symbol ('), being 1/60 of an hour, are equivalent to 1/4 of a degree. All of this seems quite complex at first, but it relates to our overall theme of ease of use: these mathematical relationships are human inventions that greatly simplify calculations. As a counterexample, the fourth-century BCE Chinese astronomer Gan De, who knew the length of the year precisely, broke with tradition by dividing the sky into 365 1/4 degrees. This was more accurate, but because it was also less convenient mathematically, it did not come into widespread use.

Now that you know the derivation of the terms, remember that a minute of angle is *not* the same as a minute of time. To avoid confusion, I will be giving measures of angles using this notation: 15° 10′, which means "fifteen degrees, ten minutes," which we can convert to 15.66°. As the ancients did, I will also give the coordinates of a celestial object by saying it is "in" a certain zodiac sign, with each of the 12 astrological signs signifying an arc of 30° in the sky.

That the earth was round, and even its size, had been long known. Cleomedes, another ancient Greek astronomer (albeit someone about whom we know even less than we do about Ptolemy), tells in his book *On the Circular Motions of the Celestial Bodies* how Eratosthenes of Cyrene, chief librarian at the Library of Alexandria, calculated the circumference of the earth in about 240 BCE.[23] Since Eratosthenes knew the distance between Alexandria and the Egyptian city of Syene, by comparing the difference in the length of the shadow of a measuring stick at the sun's zenith (highest point) on midsummer's day at the two locations, he could, by simple geometry, find the difference in the angle of the sun and calculate how far apart the two cities were on a sphere. This turned out to be 7.2°, or 1/50 of a circle. By multiplying the distance between the two cities by 50, he could then find the earth's circumference, which he calculated to be 252,000 stadia—not far off the mark, if he was using the 157.6-meter (517-foot) Egyptian stadion as his measure instead of the 185-meter (607-foot) Attic (Greek) stadion. We know today that the earth's true circumference around the poles is about 40,000 kilometers, or 24,860 miles.

The other heavenly bodies whose movements were used by ancient astronomers are the sun and the moon. (The classical planets—Mercury,

Venus, Mars, Jupiter, and Saturn—though they are significant for the history of astronomy, are less important for timekeeping.) Note that, in the Ptolemaic model, all these heavenly bodies were thought to be mounted on their own solid, invisible sphere, each of which had its own motion. Outside of the whole is an invisible final sphere, the *primum mobile* (prime mover or first moved), which imparts motion to the rest of the spheres.

The Astronomical Basis of the Year

Now that I have discussed my sources and how ancient astronomy used mathematics to measure natural cycles, we can move on to look at how ancient people used these ideas to construct calendars.

The Sun and the Year

Because most human activities take place during the daytime, the most natural way of telling time is by the sun. However, this is more complicated than it might seem. Because the earth's axis is tilted relative to the plane of its orbit, the length of the day varies over the course of the year. This is because we are seeing the sun at different angles. For instance, according to modern time reckoning, in Paris, France (located at 48° 50′ north latitude, 2° 20′ east latitude), daylight can last from over 16 hours and 10 minutes during the summer solstice to 8 hours and 15 minutes during the winter solstice. The varying length of the day and the fact that the sun rises and sets in a different place throughout the year are the most noticeable phenomena associated with the change of the seasons. Note that the longest day of the year is the *summer solstice*, while the shortest is the *winter solstice*; the days when light and darkness are equal in length are the spring and fall *equinoxes*. The time between equinox and equinox, or solstice and solstice, is called the *tropical*, or *solar*, *year*.

Again, remember the modern hours that I just used to describe the length of day are equivalent to 1/24 of a mean solar day, that is, an imaginary day of "average" length. This is a modern concept, and it required more sophisticated observations than were possible in the premodern world, which is why the sidereal day was primary. We'll look at the development of the mean solar day in chapter 4. Though the use of hours of equal length was certainly known from astronomy, because of the changing length of day and night, in premodern times, measurements were often made in *unequal hours*, which were simply divisions of the periods between sunrise and sunset and sunset and sunrise into 12 parts each, which would be longer or shorter depending on the time of the year. (That is, the daylight hours would be longer in the summer and shorter in the winter, and vice versa.) Because unequal hours were both easier to measure and more useful, they were widespread in both Europe and China.

A more subtle annual change, caused by the fact that the tilted earth is orbiting around the sun, is the sun's apparent motion against the background of the stars. Of course, we can't see the stars when the sun is in the sky, but if you were to observe the predawn sky, the last stars seen to rise would change over the course of the year. We call the apparent path of the sun in the celestial sphere the *ecliptic*, as seen in figure 1.1. At some point in the ancient world, it was noticed that the time it took the sun to travel through the ecliptic and for the rising of the stars to come back to the same point in their cycle coincided with the cycle of the year. The circle of the constellations chosen to form the zodiac, which follows the plane of the earth's rotation, is called the celestial equator. It is inclined 23.5° from the ecliptic, which is also, not coincidentally, the plane of the planets' orbits around the sun.

The northern and southern points of the ecliptic, where the sun seems to "turn" aside from its north-south wandering on the solstices, are called the *tropics* (from the Greek for "turn") and are named after the zodiac signs in which the sun seems to change its course. The northern point is called the Tropic of Cancer, the southern the Tropic of Capricorn. To someone standing on the Tropic of Cancer, the sun will seem to be directly overhead at noon on the summer solstice, while the sun will be directly overhead at the Tropic of Capricorn at the winter solstice. This is why the sun's approximately 365¼-day cycle from one tropic back to the other is called a *tropical year*.[24]

The fact that the sun moves predictably but not in unison with the rest of the celestial objects is why Ptolemy imagined it as mounted on its own invisible sphere, which moves independently of the earth and stars. The moon and the five visible planets (Mercury, Venus, Mars, Jupiter, and Saturn) were likewise considered to have their own spheres. The fact that the visible celestial bodies are seven in number is probably the reason why there are seven days of the week, each named for one of the seven "wandering stars"—*asteres planetai* in Greek, from which we get our word "planet."

The Moon and the Month

The moon is the basis of the reckoning of the month; indeed, "moon" and "month" are cognates in most languages. The most obvious observation we can make about the moon is that it changes phases. This is caused by the fact that we see the sunlit part of the moon from different angles—*not* because the earth shadows the moon (that is a lunar eclipse). If the sun and the moon are on the same side of the earth, the sunlit portion of the moon is hidden on the far side and we have a new moon; if they are on opposite sides, the sunlit portion is visible and we have a full moon. This fact was well known in the ancient world, which, even though it believed the sun and moon both went around the earth, also believed the sun was more distant than the moon. The cycle of the phases of the moon is called the *synodic month*. It lasts 29.53 days.

However, the moon, like the sun, also moves with respect to the background stars—in this case, because the moon orbits the earth. The *sidereal month* is the time it takes for the moon to return to a given position against the stars. Its duration is 27.32 days. There are other types of months as well, but the synodic and sidereal months are the most important for our purposes.

The difference between the synodic and sidereal month is caused by the fact that the moon is orbiting the earth at the same time the earth-moon system is orbiting the sun. The moon thus has to travel a bit more than 360 degrees in its orbit before it will be in the position where the sun will again illuminate it as a full moon. Thus, though the moon orbits the earth once a sidereal month, the phases of the moon do not align with this number. The core of the problem of constructing a calendar is that the number of true sidereal months in a solar year is an irregular number.

The difference between the synodic and sidereal month makes constructing a system that accurately keeps track of both the sun and the moon rather complicated. Ancient civilizations came up with a number of ways to solve this problem. One of the functions of the Antikythera Mechanism was to calculate the *Metonic calendar*. The Metonic calendar is the lowest common product of the solar year—that is, the amount of time it takes the sun to complete its journey through the seasons, from spring equinox to spring equinox or summer solstice to summer solstice—and the synodic month—that is, the period of the moon's phases. The Metonic year comes to about 19 solar years or 235 months. This aided cultures that primarily used lunar calendars, such as the Mesopotamians, Arabs, Greeks, and Hebrews, to add leap months at regular intervals and thus keep their calendars in phase with the cycle of the year. (See the sidebar for some ways in which different cultures kept calendars.)[25]

The Roman Calendar

The modern Western secular calendar—the one kept the world over—originates from the Roman calendar. To understand its development, you need to understand how increased precision of astronomical observation led to its becoming the calendar we are familiar with today.

Originally, the Romans kept a 10-month calendar, beginning in the spring equinox with March (Martius), named after Mars, the god of war, since that was when the military campaign season began. March was followed by April (Aprilis), possibly named after the "opening" of flowers; May (Maius), the "elder" month; and June (Iunius), the "younger" month. The remaining months were simply numbered "fifth" through "tenth"—Quintilis, Sextilis, September, October, November, and December. At some point, two more months, Ianuarius (after the god of doors) and Februarius (named after a ceremony of purification) were added to the beginning of the year—though the seventh month remained "Quintilis." This brought the calendar roughly into accord with the number of lunar cycles per year but is why even though October means "the eighth month," it is actually the tenth month of the year, "November," the "ninth month," is the eleventh, and December, the "tenth month," is

Calendars and Cultures

Although the astronomical basis of timekeeping doesn't change, the irregular cycles of natural phenomena present a number of challenges for making a comprehensible calendar. The fact that we use the solution the Romans adopted is only a historical accident; accordingly, the ways in which other cultures kept calendars and recorded dates varied widely. Here is a brief synopsis of some of the chief calendar systems of the ancient world.

The Chinese Calendar

The Chinese calendar[1] was tremendously influential in East Asia, and it was adopted by the Koreans, the Vietnamese, and the Japanese, among other peoples. It is a lunisolar calendar, maintaining separate lunar and solar cycles. The first of the lunar month is defined as the new moon, while the solar calendar breaks the year down into twelve months of 30 or 31 days. (The Chinese also have their own cycle of 12 zodiac symbols, which they developed independently of the Greco-Babylonian tradition.) Because the synodic month and the year don't match up, a thirteenth "leap month" is added every two or three years, though initially they were added on an ad hoc basis. The months are named for natural events: the second month is Xìngyuè, Apricot Month; and the sixth Héyuè, Lotus Flower Month (the exceptions being the seventh, Qiǎoyuè, devoted to domestic skills; the tenth Yángyuè, Yang Month;

the twelfth. However, the custom of reckoning the year as beginning in March lasted through the Middle Ages in some places. Thus, if a certain writer gave the date of February 10, 1409, we would actually reckon it as 1410 by our calendar.

Roman systems of counting the days were quite complicated, as well. All the months, save February, the month of purification, had an odd number of days, because even numbers were considered unlucky. The week had eight days, with the last being a market day. This avoided the prohibition on even numbers because the Romans counted inclusively; thus the eight-day week was called a "nine-day," or *nundium*. The seven-day week was introduced in the first century CE, and, in Romance languages, the days still bear the names of the Roman gods, who were also

the eleventh Dōngyuè, Winter Month; and the twelfth Làyuè, Preserved Foods Month). The Chinese also use 10- and 7-day weeks (the latter of which was likely introduced from outside China).

Beginning in about 600 BCE, the Chinese began reckoning a unit of time called the *qi*, which has no counterpart in the west. A qi was defined as when the solar ecliptic longitude moved 15°, and a solar year had 24 qi and 5.24 extra days. To reconcile this, an intercalary day was added every 70 days or so. The practice of adding a "leap" day, week, month, or other period of time to reconcile calendars to natural cycles is called *intercalation*. This means that Chinese astronomers took meticulous observations of the sun, and the astronomer Zhang Zixin noticed in the sixth century CE that the sun's speed through the celestial sphere varies throughout the year—a very sophisticated observation.

The Jewish Calendar

The Jewish calendar is, like the Mesopotamian, based on the lunar cycle. Because it is a living tradition, still used for the Jewish liturgical year, it has changed quite a bit from biblical times. In the first centuries CE, when Palestine was under Roman rule, the months were reckoned from sighting the crescent moon, with corrections made as needed so that the months, and especially the spring month of Nisan in which Passover was celebrated, stayed in line with the cycles of the agricultural year. To do this, religious authorities needed to confirm not just the astronomical

(*continued*)

the classical planets: the days "of the moon" (Lunae), of Mars (Martis), Mercury (Mercurii), Jupiter (Iovis), Venus (Veneris), and Saturn (Saturni); only Sunday (*dies Solis*) was renamed the "Lord's day" (*domino*) after Christianization. (English and German days of the week are named after Germanic gods identified with the Greco-Roman gods.)

Besides the days of the week, religious holidays also took place at certain points of the month: the *kalends* was the first of the month, and the *ides* was roughly the middle of the month, the 13th or 15th (just so long as there were 16 days until the next kalends), and *nones* was the ninth day before the *kalends*. (Thus, Julius Caesar was assassinated on March 15.) Days were also given a certain religious character—for instance, they could be *fasti*, on which public business could be conducted, *nefasti*, when it could not, and *feria*, public holidays, which carried over into medieval

equinox but also the ripening of fruit and grain. After the Disapora, or scattering, of the Jewish people to places where such signs might not sync up with those in Palestine, it was replaced by mathematical calculations that were summed up and codified by Maimonides in the twelfth century CE.[2] (In this reliance on expert calculations, Judaism differed from Islam and was more in line with Christianity.)

The lunar months in the Jewish calendar are alternately given 29 or 30 days. As with other calendars, the Jewish calendar reconciles the solar and lunar years by doubling the month of Adar according to a prescribed schedule based on the 19-year Metonic cycle. The intercalation of the Jewish calendar is not exact and still leads to a one-day discrepancy with the tropical year every 216 years.

Judaism, of course, uses a seven-day week corresponding to the biblical story of creation, while the Jewish date counts from the year of creation as calculated by Rabbi Jose ben Halafta in the second century CE and later confirmed by Maimonides (rendered AM, or *anno mundi*, year of the world). Prior to this, the Jewish people had used the Greek calendar, which was widespread in the ancient Mediterranean following the conquests of Alexander the Great. The Jewish New Year is Rosh Hashanah (literally the "head of the year"); this begins the Ten Days of Atonement, which culminate in the solemn holiday of Yom Kippur. (Interestingly, Rosh Hashanah begins the Jewish civil calendar, but Nisan begins the religious calendar.) However, finding the date of Rosh Hashanah is not as simple as merely consulting a calendar: there is a complicated series of rules concerning the phase of the moon and the day of the week on which the holiday falls that is consulted to deter-

Christianity as weekdays when believers had to attend mass. Since medieval markets were held on such days, our word "fair" comes from *feria*.

Likewise, the Romans had a number of systems for the year: *ab urbe condita* ("from the founding of the city," abbreviated AUC), which supposedly took place in 753 BCE, was one system. However, AUC was not used before the late Republic; earlier, it was more usual to give the year by referring to who were the consuls, or chief executives of the state. Since we have a complete list of consuls, we can date events in Roman history relatively accurately. Other unwieldy Roman measures of the year included a 15-year cycle called an *indiction*, as well as later the regmine whether the adjustable months in the calendar should be length-ened to postpone Rosh Hashanah on the lunar calendar.

The Islamic calendar

The Islamic calendar is simpler than those of the other Abrahamic religions, Judaism and Christianity: it is a strictly lunar calendar of 354 or 355 days. This means that it lags behind the Christian calendar by 10 cumulative days per year and that the cycle of months will shift relative to one another in a cycle that takes 33 years to complete.

The prophet Mohammad repeatedly emphasized that religious knowledge should be accessible and known to all believers. As part of this, he forbade intercalary months in the Quran.[3] In his farewell sermon, which was passed down orally and preserved by early historians such as Muhammad ibn Isḥāq and Abū Jaʿfar Muhammad ibn Jarīr al-Ṭabarī, the Prophet said:

> So beware of him in your religion, O people, intercalating a month is an increase in unbelief whereby the unbelievers go astray; one year they make it profane, and hallow it another, [in order] to agree with the number that God has hallowed, and so profane what God has hallowed, and hallow what God has made profane. Time has completed its cycle [and is] as it was on the day that God created the heavens and the earth. The number of the months with God is twelve: [they were] in the Book of God on the day He created the heavens and the earth. Four of them are sacred ...[4]

The first of the month in the Islamic calendar traditionally dates from the actual sighting of the crescent moon, which can vary from

(continued)

nal years of rulers such as the *Diocletian year*, named for the Roman Emperor Diocletian, who ruled from 284 to 305 CE and was widely dis-liked by early Christian writers for his persecution of the faithful.

The problem with the Roman calendar was that it had only 355 days, which was solved by putting in an intercalary month of 22 or 23 days after the first 23 days of February. Ideally, this was done every other year, giving an average of a 366 1/4-day year. (The extra day could be corrected by further refinements.) However, the decision to change the calendar was made by magistrates, who, as part of the general break-down of public order and grasping for power in the late Republic, often

place to place owing to the earth's position and weather conditions, while the Islamic year dates from the *anno hegirae* (AH), the year of Mohammad's flight from Mecca to Medina in 622 CE.

Calendars of the Indus Valley

Archaeologists have excavated tally sticks from Harappan civilization (which flourished between 7,000 and 2,000 BCE) that indicate some sort of calendrical system based on astronomical observation.[5] However, what this system was is difficult to determine. It is not until the first centuries BCE that we had a definitive written source, the *Vedānga Jyotiṣa*.[6]

The traditional Hindu calendar is not too different from the Jewish or Mesopotamian calendars in that it is lunisolar. Like the ancient Jewish calendar, it adds an extra lunar month every few years, based on observation of natural phenomena. This should not be surprising, as the subcontinent was connected to the Middle East via trade. Indian scholars also had highly sophisticated mathematics and made detailed observations of the sun, stars, and moon. They were influenced in this by the Greek astronomical tradition and in turn influenced Arabic, Western, and Chinese astronomy.[7]

The Mesoamerican Calendar

The calendars of pre-Columbian Mesoamerica[8] came about entirely independently of the Old World and form an integral tradition that

decided to extend years in which they or their allies were in power. The situation got so bad that the Roman calendar ultimately wound up 67 days off from the tropical year!

The civil wars of the first century BCE ended with Julius Caesar becoming the dictator of Rome. He had learned about Greco-Egyptian timekeeping while conquering that country and, advised by scholars such as Sosigenes of Alexandria, decided to set matters aright. First he realigned the Roman calendar and the tropical year by adding two intercalary months between November and December, then he abolished intercalary months altogether. Instead, he brought the calendar up to 365 days by adding two days to January, Sextilis (August), and December; and one to April, June, September, and November; and a leap year every four years.

lasted from the Olmec and Zapotec cultures around 1800 BCE to the European conquest in the sixteenth century CE—and are, in fact, still partially used by the descendants of the Maya in the highlands of Mexico. They are characterized by a base-20 system and the use of zero as a placeholder.

For day-to-day use, the pre-Columbian Mesoamericans used a 260-day cycle (known as a *haab'* in Mayan and *tonalpohualli* in Aztec) and a 365-day cycle (known as a *tzolk'in* in Mayan and *xiuhpohualli* in Aztec) that roughly corresponded to the solar year. The lowest common multiple is 18,980 days, or about 52 years, which was called the Calendar Round. They also used a series of glyphs, or symbols, to denote the phase of the moon. For longer periods, such as for monuments and histories, they used the Long Count Calendar. Since the Mayan calendar is probably the best-known example of this calendar system, we will use the Mayan terms.

To understand the Long Count, you must understand that it combines both base-20 and base-18 counting. The Mayan name for a day is a *k'in*; 20 k'in made a *winal*, and 18 winal (360 days) made a *tun*. Twenty tun (7,200 days, or about 19 years and 260 days by the Gregorian calendar) make a *k'atun*, and 20 k'atun made a *b'ak'tun* (about 394 1/2 years). The count begins on the mythical creation date of August 11, 3114 BCE. December 21, 2012, marked the beginning of a new b'ak'tun cycle, but this was nowhere near the prophesized end of the world, as some people believed—it was simply the odometer

(continued)

Month names did not change immediately, but Quintilis and Sextilis were later renamed July (Iulius) and August (Augustus) to honor Caesar and Augustus, his adopted son and successor. (The old calendar's influence was seen in one interesting way: the Julian leap year was actually done by doubling February 24, and this is still maintained in the liturgical calendar of the Roman Catholic Church.)

Early Christianity and the Roman Calendar

The Julian calendar was not the only calendar in the Roman Empire. It became standard because the Christian church adopted it—and

turning over. Note that, by the time the Spanish arrived, the "long count" had fallen out of favor, and the Maya were using a "short count" of 13 k'atuns (256 years and 160 days).

1. See Mitsuru Sôma, Kin-aki Kawabata, and Tanikawa Kiyotaka, "Units of Time in Ancient China and Japan," *Publications of the Astronomical Society of Japan 56*, no. 5 (October 25, 2004): 887–904.

2. See, for instance, Ari Belenkiy, "Astronomy of Maimonides and Its Arabic Sources," in *Cosmology across Cultures ASP Conference Series*, vol. 409, *Proceedings of the Conference Held 8–12 September 2008*, edited by José Alberto Rubiño-Martín, Juan Antonio Belmonte, Francisco Prada, and Antxon Alberdi (San Francisco: Astronomical Society of the Pacific, 2009), 188–202. On the time of transition, see the anecdote about Rabban Gamaliel (in the first century CE) in the *Mishneh Rosh Hashanah* at https://www.sefaria.org/Mishnah_Rosh_Hashanah.2.7?lang=bi&with=all&lang2=en, accessed June 7, 2018.

3. Qurān 9:36–37.

4. Abū Ja'far Muḥammad ibn Jarīr al-Ṭabarī, *The History of al-Tabari*, vol. 9, *The Last Years of the Prophet*, translated by Ismail K. Poonawala (Albany: State University of New York Press, 1990), 112–114.

5. Walter Ashlin Fairservis, *The Harappan Civilization and Its Writing: A Model for the Decipherment of the Indus Script* (Leiden: E. J. Brill, 1992), 60–66.

6. V. N. Tripathi, "Astrology in India," in *Encyclopaedia of the History of Science, Technology, and Medicine in Non-Western Cultures*, 2nd ed., edited by Helaine Selin (New York: Springer, 2008), 264–267.

7. See S. Balachandra Rao, *Indian Astronomy: An Introduction* (Hyderabad: Universities Press, 2000).

8. See Prudence M. Rice, *Maya Calendar Origins: Monuments, Mythistory, and the Materialization of Time* (Austin: University of Texas Press, 2007).

anywhere Christianity went, so, too, did the calendar of religious observances. Though many earlier emperors had persecuted Christians, the Emperor Constantine officially tolerated the religion in 313, and the church rapidly gained in power and influence in the Roman Empire. (All dates hereafter are, unless otherwise noted, CE.) Constantine also established a new, eastern, capital for the empire at Constantinople, which is modern Istanbul. (Constantinople had formerly been called Byzantium, which is how the Byzantine Empire got its name.) In 380, the Emperor Theodosius, the last emperor to reign over both the eastern and western halves of the empire, made Nicene Christianity—that is, Christianity that follows the statement of faith laid down by the council Constantine called at Nicaea in modern Turkey in 325—the official religion. During the period of Rome's breakup, bishops were some of the only authorities left in the world, and great thinkers such as Augustine of Hippo laid the foundations by which Christian society would be run thereafter. Likewise, Benedict of Nursia (c. 480s–540s) established influential rules for those who wished to live holy lives of prayer and contemplation as monks.

Another reason why the Christian calendar became widespread in Europe was because of Charles the Great (740s–814), who we call Charlemagne after his Latin appellation, Carolus Magnus. The Germanic peoples who had carved up the Roman Empire among themselves were by and large pagans, but they slowly began to convert—or be converted—to Christianity. Most important among them were Charlemagne's people, the Franks, who had settled in Gaul (roughly modern France). On Christmas Day of 800, Charlemagne was crowned Emperor by Leo III, bishop of Rome. Of course, there *was* no longer a Western Roman Empire to rule; what this meant was that the axis of politics in the West would be between the power of a secular ruler and the authority of the Roman Church, led by the bishop of Rome—that is, the pope. Religion was a tool of governance: churchmen, being literate, served as administrators, and to be under Frankish rulership meant that you were forced to accept Roman Christianity and to properly observe all of its rituals and beliefs.

Charlemagne and his heirs were also great patrons of schools and scholars. In fact, we have many of the books that have survived from the ancient world only because they were recopied into a new, much more legible handwriting by Carolingian scribes. All of this was done because, to Charlemagne and his successors, proper learning—including maintaining the calendar—was proper religious observance and, therefore, the proper ordering of this world. Charlemagne also renamed the Roman months in German after activities in the agricultural cycle, and these names remained in use in some German-speaking lands hundreds of years after his death. However, alongside this was the Christian liturgical calendar, which all people had to abide by—unless they wanted to be persecuted as heretics. Medieval civilization was based on this idea of church and crown working together to discipline and rule society.

The Stars and the Hour

To be a good Christian, it wasn't enough just to know the right *date* for religious rituals: it was also necessary to know the proper *time of day* to pray. Christianity thus had an enormous influence on Western timekeeping, because knowing the time for prayer—which meant watching the stars and other natural signs—was a religious duty. The word *hora* (hour) occurs in the Latin (Vulgate) Bible in 152 places, 35 of which mention specific hours or the specific measurement of time to be adopted as the times for prayer. Many of these passages, such as the parable of the workers in the vineyard from Matthew 20:1–16, were well known and often quoted in sermons and theological writing. The Roman system of hours thus became an indispensable part of the Christian world.

Before Rome had become a great empire, the Romans were not actually very interested in sundials or accurate timekeeping but only the practical considerations of marking events. For instance, for lawyers' arguments to be valid in Roman courts, they had to be made before noon, so when the sun was sighted between two points in the Forum, it was too late to plead. (The Roman emphasis on times before and after noon give us AM and PM, for *ante meridiem* and *post meridiem, meridies* being the Latin for "noon.") The early Roman indifference to accurate timekeeping is shown by an anecdote given by the later writer Pliny the Elder: in 263 BCE, the general Valerius Messalla came back to Rome with a sundial he had captured as war booty from the Greek colony of Catana (modern Catania) in Sicily. However, because of the difference in latitudes, the sundial didn't give accurate time for Rome—and, for almost a century, the Romans didn't seem to notice or mind.[26]

This began to change as the Romans came into contact with Greek science. In 164 BCE, they acquired a sundial calculated for their latitude. They soon adopted the Greeks' twelve-hour unequal divisions of day and night, as well as four "watches" of the night, and sundials became status symbols. Sundials were built into Roman public buildings, such the ornate Tower of the Winds in the Roman Agora in Athens, which also had a weathervane and a water clock.[27] Still, the Romans were not greatly interested in accurate timekeeping, never felt the need to use equal hours in everyday affairs, and certainly never felt the urge to automate the process. Although some effects of some Roman timekeeping customs remain today, such as using AM and PM, the custom of reckoning the day as beginning at midnight, and the Spanish *siesta*, which comes from the Latin "sixth" (or *sextus*) hour—halfway through the day, when the sun was at its hottest—the Romans' most important legacy was to pass on Greek and Mesopotamian ideas in a practical way. (An example is seen in the book *De architectura* by the Roman writer Vitruvius; I have included his instructions on drawing a sundial as one of the practical exercises at the end of the book.)

Because of the cycle of prayer, Christian monks of the Middle Ages were much more concerned with the time than the Romans had been. The Benedictine Rule, in keeping with Psalm 119:62 ("at midnight I rise to give thee thanks") and 164 ("seven times a day I praise thee"), established their eight-times-daily round of prayer: Matins (sunrise), Sext (midday), Compline (sunset), and Laudes (around midnight), to which were added the Roman quarter-hours mentioned in the Bible, whose timing depended on the natural signals: Prime (shortly after Matins); Tierce (later in the morning), None (which was originally midafternoon but gradually moved closer to modern noon, to which it gives its name, over the course of the thirteenth century), and Vespers (around sunset).

It is important to understand that the medieval "hours" were not as precise as our modern schedules, where we *always* have to be school or work at 9 AM. Rather, they floated through the year. For instance, chapter 8 of the Benedictine Rule specifies that during winter, that is from November to Easter, monks should arise in the "eighth hour of the night, so that they will arise having rested a little more than half the night and arise refreshed"—in other words, the "eighth hour" is by definition "a little more than half the night."[28] In other times of the year, however, prayers were arranged so that Matins could be said a short time later at daybreak, after the monks have had time for "necessities of nature." Times for work and prayer were specified, as well—in the 48th chapter of the Rule, the office of None is at the "middle of the eighth hour" from Easter to October, and, from October to Lent, Tierce is at "second hour." (As specified in the various monastic rules, the bells rang for the offices, not the hour.)

The Hours in Premodern China

As scientific observations in historical records reveal, the Chinese also had a sophisticated and astronomically based system for marking both equal and unequal hours, one that was quite different from the Western solution.[1] First, the day from midnight to midnight was divided into 100 equal parts (*ke*, in the modern Pinyin system of transliteration). Each ke is equal to 4.4 minutes, or 14 minutes 24 seconds, in Western reckoning. This system was first recorded in the *Hanshu*, compiled in 82 CE by the historian Ban Gu of the Han dynasty and lasted until 1628, when the Chongzhen emperor, the last ruler of the Ming dynasty, made the day 96 ke long, with each ke lasting 15 minutes. Second, Emperor Ping of the Han dynasty (r. 9 BCE–6 CE) instituted a system of twenty-four equal hours, which were incorporated into twelve double hours (*zhi*); each double hour lasted 8 1/3 of the older ke. These twelve hours were named for animals corresponding to the Chinese calendar: rat, ox, tiger, and so on. The Chinese used sundials and water clocks to keep track of the hours, with readings made in intervals of 2.4 minutes (the common denominator of the two systems).

In addition to this rationalized timekeeping, the Chinese also had the *geng* system of unequal hours, which divided the day from dusk to dawn into five parts, as well as an urban signal system based on beating a drum or gong. The *Engishiki*, a tenth-century book of regulations, gives these indicators in precise astronomically based times given in ke and zhi. In other words, looking at the historical record, Chinese timekeeping was no less rationalized than Western, and, in fact, it was a great deal more sophisticated at an earlier date. The big difference was not one of thinking or organization but (as we'll see in the next chapter) the development of increasingly complex mechanical devices to automate the process.

1. Adapted from Mitsuru Sôma, Kin-aki Kawabata, and Tanikawa Kiyotaka, "Units of Time in Ancient China and Japan," *Publications of the Astronomical Society of Japan* 56, no. 5 (October 25, 2004): 887–904.

Using the Stars to Tell Time

So how did monks use the stars to determine the times for work or prayer? There are actually three answers to this question. The first, and simplest, is that they didn't commonly determine the time with our modern hours of 60 equal minutes, but rather they used the unequal or seasonal hours—which is why the times floated so much. During the course of any given night, six zodiac signs will rise and set—though, of course, *which* six will vary through the year. Each of the six will take, by definition, two seasonal hours to rise, so that, by gauging the march of the zodiacal signs across the sky, we can know what proportion of the night has passed. Of course, the earth's rotation does not speed up in the summer; rather, because the angle of the earth relative to the ecliptic changes, the zodiac signs have more or less of an arc to traverse from the point of view of an astronomer at a particular latitude. The seasonal hours are thus proportionately shorter at night and longer during the day in the summer, when the nights are shorter, than they are in the winter, when they are longer at night and shorter in the day.

These simple astronomical cues seem to have been the most popular method of determining the hour in the early Middle Ages. St. John Cassian (360–435), who founded a monastery at Marseilles in southern France and wrote an influential book called *On the Institutions of Monasticism* (*De coenobiorum institutis*), specifies watching the stars to determine the time for prayers, while Gregory of Tours (538–594), who was bishop of the city of Tours in France's Loire Valley, and Bede (672–735) give detailed instructions on taking astronomical observations to know when to sing the night office.[29] These cues were not as sophisticated as Ptolemy's astronomy—early medieval observers were merely looking for when a certain star reaches a certain place—but they were a remnant of the classical legacy.

The second, more complicated way of telling time from the stars is to find the equal hours using mathematics. In fact, astronomy and time-keeping—to determine the times of the rising and the setting of the zodiac signs in any season—was one reason why the Mesopotamians and Greeks developed trigonometry.[30] To calculate even hours, we need to draw a coordinate system on the sphere of the stars. The simplest way of doing so, as I mentioned before, is to project the earth's system of latitude and longitude onto the heavenly sphere. The sphere of the stars, like the earth, has a north pole and an equator. We can measure along an imaginary line drawn parallel to the celestial equator and so find the *right ascension*, or "longitude," of any star in the sky. The band of the zodiac signs also follows this "longitude" and, as noted above, was historically used to describe the positions of celestial bodies. A related concept is the *oblique ascension*, which is the altitude of an imaginary point on the celestial equator that rises at the same time as our star and was more used in ancient astronomy. In the same way, we can measure the *declination*, which is equivalent to latitude—the distance above or below the celestial equator, measured at a 90-degree angle.

This would be all well and good if the earth were not tilted on its axis and we were standing at the equator: because the heavenly sphere moves 15 degrees per hour per day, we could simply use a sighting device to measure the altitude of a particular star and thereby tell the time. But because the earth is tilted on its axis, the star will be traveling at some angle relative to the horizon, parallel to the celestial equator and at an angle to the ecliptic. (If this seems complicated, remember that, put another way, all our chosen star represents is a point on the "face" of the celestial clock, which rotates around the polestar 15 degrees per hour, but the altitude of the polestar will vary according to the year.) Using trigonometry, we can correct for this. However, because spherical trigonometry is beyond the scope of this book (though I'll reference it again when I discuss astrolabes in the next chapter), if you wish to study this more in detail, I recommend taking a course on astronomy.

The third method of keeping time by the stars makes the second way simple again: we can look them up in a book. Ptolemy included the figures of right and oblique ascensions for select stars in an appendix to the *Almagest*, most commonly called the *Handy Tables*. Using these, an astronomer can easily find the sidereal time—provided the table is computed for the correct latitude! Tables of ascensions have been used in books of astronomy for thousands of years and tell us when celestial signs will rise—or if we sight a certain star and measure its altitude, what time it is.

For medieval Christians, as for the ancient Babylonians and for traditional Chinese society, gleaning information from the motion of the sky, or modeling it, wasn't simply a religious duty or abstract knowledge: it was a glimpse of the sublime. This was another part of the ancient world's legacy to the Middle Ages: for instance, in the section known as the "Dream of Scipio" from the first-century BCE Roman orator and politician Marcus Tullius Cicero's *De re publica* (*On the Republic*), the Roman general and hero Scipio Aemilianus is raptured into the heavens to observe the sublime motion of the heavenly spheres, hear their subtle music, and to meet his father and grandfather, who tell him that the virtuous have their reward among the stars. The pagan philosopher Macrobius (fl. 395–423) wrote an astronomical commentary on this passage that was widely read in the Middle Ages. Why was this pagan philosopher's work so important to a Christian society? To Macrobius, the spheres are not only natural phenomena but also manifestations of divine will. His commentary is thus not only a précis of ancient astronomy but also a primer of ancient philosophy, teaching that while earthly things are to be held in contempt, the virtuous have their reward in heaven. Because the spheres, like music, may be understood mathematically—the science of number and proportion—later Christians understood this to mean that humans, created in God's image, can take their place among the stars—"have their reward in heaven," as we still say. It follows that knowledge of the regular motion of the stars, timekeeping, and godliness go hand-in-hand. So, by observation, one could even discern the hidden properties of creation and, from an elevated perspective, discern more of the divine plan than other mortals—God's will written in the cosmos, written in number.

The wise use of time was also a longstanding trope of virtue throughout the Middle Ages. The fifth book of the seventh-century Spanish bishop Isidore of Seville's famous encyclopedia, the *Etymologiae* (*Etymologies*), is on "laws and time"; the two subjects are the regulators of the world. In the same way, candles were of religious, symbolic, and practical importance, and marking time by the burning of candles is a trope that occurs often in the lives of saintly kings. The Welsh monk Asser, in his biography of the English ruler Alfred the Great (849–899), records the king as having six candles, each of 12-pence weight, burned every day in a specially constructed horn lantern.[31] Each candle, in turn, was marked by 12 divisions. The idea here was obviously to mirror the twelve-hour cycle of the equal hours, but there would have been no way of knowing that exactly 20 minutes had passed for each division of a candle; Alfred was in effect proportionately dividing his day.

Later, timekeeping devices such as clocks and astrolabes also acquired moral meanings: Abelard, writing to Heloise on furnishing her convent in 1136, mentions a water clock as a necessary and ordinary part of the furnishings of a church.[32] (Heloise also notably named their son Astrolabe.) A psalter from the beginning of the twelfth century attributed to Blanche of Castile (1188–1252), mother of King Louis IX, shows on the verso side of its first folio a miniature of an astronomer taking measurements of the stars with an astrolabe while one assistant reads from a book in Latin and another writes in a Latin manuscript—an appropriate subject for a psalter, considering that prayer times were set by observation.[33] The thirteenth-century French bishop Guillaume Durand, in his *Explanation of the Divine Offices* (*Rationale divinorum officiorum*), a handbook of church furnishings and their symbolic meanings written sometime before 1286, says that the clock teaches the priests to mind the canonical hours according to Palm 119.[34]

So, it is important to remember that timekeeping is not, and has never been, just knowing "what time it is." In both the ancient and the modern worlds, timekeeping is not only a means of regulating society but a means of predicting the future and understanding the natural universe, and it has moral significance, as well. This is how timekeeping began in medieval Europe—monks and priests measuring the sky to determine the times for prayer and bring earthly activity into line with divine will. But such timekeeping wasn't what we would call precise or scientific. Measuring durations, such as the unequal hours, was far more important than precisely and accurately "telling time" in the modern sense (that is, mirroring the revolution of the stars). In the next chapter, we'll look at how a special sort of sky-measuring device—the mechanical clock—was introduced in the Middle Ages, and how the increased precision and accuracy it allowed for changed European society, beginning the shift from comparing relative durations to measuring equal hours—with a resulting shift in mentality.

[TWO]

Cities and Clocks

Frère Jacques
Frère Jacques
Dormez-vous?
Dormez-vous?
Sonnez les matines
Sonnez les matines
Ding-dong-ding
—Traditional

IMAGINE TAKING an evening stroll through the cramped, zigzagging streets of fourteenth-century Paris. Think of how different it is from the modern city—automobiles and bicycles replaced with horses, the muddy streets paved with planks rather than cobblestones, the merchants beginning to shutter the open-front shops that occupy the ground floors of their half-timbered houses. Consider the crowding—Paris was a town of perhaps 250,000 souls crammed into the acreage of a medium-sized college campus. Think of the different smells—the aromas of cooking dinner, wood smoke, incense, and perfumes mixed with odor of unwashed bodies, rotting garbage, and human and animal waste. Then imagine the sounds—not just conversations in an almost-familiar French, the clip-clop of hooves, and snatches of laughter and song coming from open tavern doors but also the sound of church bells ringing in a complicated code that the people of Paris knew how to interpret as a language unto itself.

For instance, though the echoes of the bells of Notre Dame ringing Vespers might already have been fading into the dusk, the leatherworkers' apprentices, as per the rules of their guild, could not quit work and close their shops until the rector of their parish church rang his own bell. However, their fellow apprentices in the mailmakers' shops, dependent on their masters' whims and not the bell system, could continue to work by candlelight, glumly piercing rivet holes in the ends of the curled and flattened pieces of iron wire that would be linked together into armor for the knights of the royal court. Meanwhile, the professors at the University of Paris listened for the bells of St. Jacques or the Carmelite monks sounding their evening prayers to know when to stop lecturing.

Churches were foremost among the signs medieval people looked to in order to navigate their paths through time and space—not only the monolithic church as an institution but also the individual parish churches that they interacted with on a daily basis. Rather than relying on posted street names as we do today, one means of navigating the cramped alleyways was to sight the various bell towers over the roofs of the wood-frame buildings. Similarly, one might identify one's dwelling place to a notary or official by using the local parish church as a convenient landmark.

Besides their use as spiritual and geographical signposts, these church steeples also acted as time markers. In addition to indicators of "natural" time such as the passage of the sun through the sky, the rhythm of daily life was regulated by the ringing of their bells according to the canonical—that is, unequal—hours.[1] Everything from the bustling marketplaces to working hours was measured by these signals. To the student, it was the signal that he was late to class; for the worker, it was the cue to begin her daily labors; for the monk or nun, it was the summons to divine service.[2] As alien as the bell system might seem to our eyes and ears, and as haphazard as some of the means of measurement might seem to us today (for instance, it was considered dark when one "couldn't tell the difference between a dog and a wolf"), it was a system of telling time no less systematic than that of our modern atomic-synchronized clocks, for it accomplished one of the major purposes for which we use timekeeping: social coordination.

To understand this system, we must first understand its context. Christianity provided a shared symbolic language and culture, a social support network, and a way of reconciling differing interests in this world while preparing for the next. By extension, the church's bell tower ringing the canonical hours did not have one sole interpretation but rather many possible different meanings, depending on the hearer and context. All bells did not even sound the same: Guillaume Durand, in his *Explanation*, specifies the six types of bells used in a church: the *squilla* (handbell), the *cymbalum* (in the refectory), the *nola* (in the cloister), the *nolula* or double *campana* (in the clock), the *signum* (in the tower), and the *campana* (in the bell tower). He also specifies that bells were to be rung 12 times a day—one for Prime, thrice for Tierce, Sext, and None, and multiple times for Vespers and Matins.[3]

Besides being a call to prayer, bells were a universally recognized symbol of danger, the air-raid klaxon of the Middle Ages. For instance, in the spring of 1278, when some of the students and teachers of the university came into the fields outside of Paris "for the cause of recreation," Gerard, the abbot of the nearby monastery of St. Germain, had his bells rung to assemble his private army, and, with cries of *Kill the clerks! kill them, kill them!* they set upon the hapless scholars, causing them to flee to the safety of the city.[4] Bells also rang for other events, such as a death, a procession, or a coronation. They could produce a variety of tones, such as being tolled slowly, rather than rung, during Lent.

Finally, bells were symbolic of the city itself. An example: in March 1331, King Philip VI abolished the city government of Laon to punish a rebellion and ordered that the bells of that city be seized, never to be returned to their towers. Only two were to remain: a large one to sound curfew, daybreak, and signal the town guard to assemble in its accustomed place; and a small one to sound a bit before the large one (the pre-signal or *clinket*, a common ringing practice) and to be rung together with it in case of fire.[5] This action speaks not only to the practical utility of such bells but also again to their social meaning—the bells were identified with the city, and to remove them was a symbolic decapitation of the community, a removal of the right to self-government and to constitute its own rules.

But how did Frère Jacques know *when* to ring his bells? As we saw in the previous chapter, the heavens were the first clock, and the unequal hours floated throughout the year. Timekeeping devices, though they would have far-reaching effects for science and society, began as nothing more than observational aids to help medieval timekeepers "read" the sun and stars. In time, such devices could be made more sophisticated, until the all-mechanical clock capable of striking the equal hours—that is, acting as a model of the heavens—made its debut around the year 1300.[6] Within two generations, a public clock that struck twenty-four-hour time would become commonplace in European cities. Urban life increasingly came to be regulated by this mechanical device, with great consequences for all realms of human endeavor. Most notably, what medieval philosophers thought about the nature of time itself was influenced by the abstract time indicated by the mechanical clock—a development that had a great effect on the development of Western science.

Early Medieval Timekeeping Technology

The Latin term *horologium*, which Vitruvius uses for his sundial, comes from the Greek *horologion* (hour counter). Eventually, it came to mean "clock"—as examples, the modern French word *horloge* or Italian *orologo*. However, in the early Middle Ages, it could mean any observational device. For instance, Pacificus, archbishop of Verona (d. 844), constructed a "night horologium" (*horologium nocturnum*), which was probably an observation tube with a cruciform sighting device or scale that could be used for measuring the ascension of stars to determine when to say prayers at night or for computing the calendar.[7]

The term "horologium" could also refer to a chart used for determining when to say prayers. For instance, the Carolingian king Pepin the Short received a horologium nocturnum from Pope Paul I (c. 758–763), along with other religious books; to judge from its context, Pepin's horologium nocturnum was not a mechanical device but a book on astronomy.[8] Another example of such a horologium is found in the *Book of Astronomy* (*Liber de astronomica*) attributed to Nimrod, a mythical king of Mesopotamia mentioned in the Bible as having built the Tower of Babel, but more likely composed during the Carolingian Renaissance of the

800s.[9] The contents include using the stars to tell time and using a duo-decimal (base-12) division of day and night, as well as astronomical tables. In one part, Nimrod's disciple Ioanton asks, "My master, how can I know what sign will be in the east and which will be coming in the west and which will be in the middle of the sky and which will be opposite it, under the earth, in summer; and in winter what the hours are of the day and night?" Nimrod responds, "Truly, my disciple, it is right that we know in all hours of the day and night what sign will rise and set. . . . [T]here-fore I will draw for you round *horologii* and around it those of the east in the west and this will show how the stars ascend and descend and in what way will be one direction or the other." Accompanying the text is a dia-gram, a horologium, showing which stars with which to reckon the time in each part of the year. The *Book of Astronomy* also shows that medieval astronomers knew about the equal hours: later in the text, Nimrod in-cludes a table giving the length of the day and night by equal hours, which vary from 15 and 9 respectively in June to 9 and 15 in December.[10]

Lastly, "horologium" could mean a device such as a sundial or water clock that could be used to determine the hour. For instance, in the biog-raphy, or *vita* (life), of St. Molingo, also known as St. Dayrgello, the seventh-century bishop of Ferns, Ireland, the saint miraculously moves, breaks, and mends a rock to make a sundial horologium in it.[11] Numer-ous simple "scratch sundials" survive in churches throughout Europe.[12] However, as we saw in the first chapter, sundials must be precisely constructed—and the Roman sundials unthinkingly copied throughout Europe were not at all accurate for telling time in latitudes north of the Alps, where the angle of the sun necessitated different proportions for the dial face. We must therefore see many of these early medieval sundi-als as what the modern historian of science Allan Mills has called "event markers"—perfectly fine for determining a conventional time to say mass but in no way suited to accurate or precise timekeeping.[13] Even when they are properly constructed, sundials also require sunshine bright enough to cast a shadow—which, if you've ever spent a winter in northern Europe, you will know is a rare thing in those lands.[14]

Thus, Europeans did not possess accurate, precise, or easy-to-use timekeeping devices in the early Middle Ages. Rather, they possessed some rudimentary astronomical textbooks and a few simple devices—a garbled transmission from the ancient world. The Arabic-speaking world was the true heritor of Greek, Persian, and Indian science, and it is from Muslim sources that accurate and precise timekeeping knowledge was reintroduced into Europe.[15]

Astronomical Timekeeping in the Muslim World

The Muslim world was as tied to regular daily prayer as Latin Christianity—if not more so, since *salat*, or five-times-daily prayer, is required of all believers. Islamic timekeeping (*mīqāt*), essential for prayer times, was predicated on the angles and lengths of shadows cast by the sun, on natural phenomena such as twilight, and on sophisticated arithmetical formulas kept by the *muwaqqit*, the professional timekeeper. Astrolabes could be used to make these readings, as could sundials. Scholars in ninth-century Baghdad read the *Almagest*, kept detailed observations, and were able to compute the exact time from their observations. More than 10,000 Arabic astronomical manuscripts survive today, and these probably represent a fraction of what once existed. Volume after volume is filled with tables of observations of the moon, stars, and planets, written in the ciphers that Muslim writers had taken from India, which would become known as Arabic numerals in the West.

Transmission of the Astrolabe

Perhaps the greatest example of the indisputable, widespread, and significant Islamic contribution to European astronomical knowledge is the astrolabe, which, from the time of its introduction into the West in about the eleventh century, proved itself useful for precisely and accurately, if not necessarily conveniently, determining the hour from astronomical phenomena as well as making other observations. An astrolabe is a form of analog computer that obviated the need to master the trigonometric disciplines required to calculate the time from the ascensions of heavenly bodies at various latitudes. A round, flat piece of metal (the *mater*) is engraved along its outer edge with the hours, degrees, or both. To this

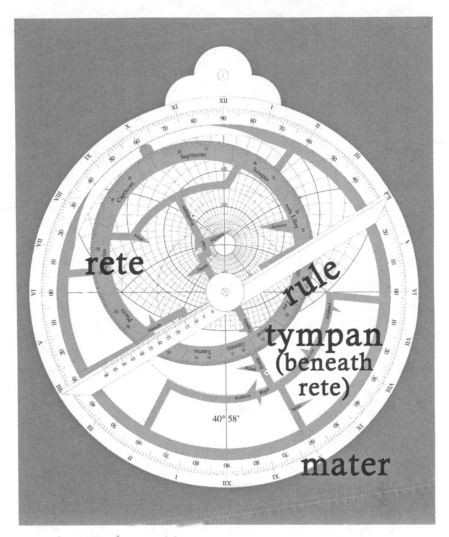

Figure 2.1. Parts of an astrolabe. Courtesy Richard Wymarc

mater is pinned a slightly smaller metal plate, the *tympan*, with a stereographic projection of the heavens and hour lines as seen from a particular latitude. Finally, a piercework *rete*, or "net," showing the ecliptic and the brightest stars to use as a celestial guide, is pinned over the tympan in such a way as to move freely. The astrolabe can be used, thanks to the *alidade*, a movable sighting rule affixed to the backside of the mater, to determine the altitude of a celestial object in degrees.

By using the alidade and aligning the rete, mater, and tympan and then reading along a rule pinned to the front side, the astrolabe can be used for determining the local time.

As with other Arabic astronomical knowledge, ample evidence exists of the penetration of the astrolabe into the West at quite an early date and its synthesis into the existing body of knowledge. Intermediaries such as the Jews living in the multicultural communities of Iberia and the south of France made such knowledge available almost as soon as it was produced.[16] As an example of this, a learned treatise attributed to Gerbert of Aurillac (955–1003) not only gives instruction on using the astrolabe but also describes the earth as divided into climate bands—an idea that was not found in Arabic sources, though it was widespread in the West.[17] Aurillac is in south-central France, and Gerbert would have lived close to the centers of translation. This knowledge spread quickly and widely from such centers: Hermann, a monk from the Benedictine abbey of Reichenau, on Lake Constance in southern Germany, wrote the first known Latin treatise on how to actually construct an astrolabe in 1045.[18] You can construct and practice using your own astrolabe in the chapter exercise at the end of this book.

Means of Measuring Duration

Measuring shadows and taking astronomical observations with astrolabes were two ways people in the premodern world could tell the time by using equal hours. However, these methods were not necessarily convenient, easy to use, precise, or accurate. Medieval people therefore used a variety of simpler means to measure duration, which they were by and large more concerned with than they were with absolute measurement. Using candles, as Alfred the Great had done, was one possibility. Guillaume de Saint-Pathus, confessor of Margaret of Provence, wife of Louis IX of France (1214–1270), details the saintly king regulating his sleep and times for reflection and prayer by the burning of a candle,[19] and Christine de Pisan, in her biography of Charles V (r. 1338–1380), notes that the king burned candles to divide the day into three parts.[20] As the daughter of the court astrologer, Christine would have been especially sensitive to the king's use of time. (Interestingly, as I will discuss below, Charles V also had the first public clocks built in Paris, but Christine does not mention this.)

Ordinary people used other indicators to measure shorter durations. The most common was the time it took to say common prayers. The anonymous author of the late fourteenth-century *Goodman of Paris*, who wrote to instruct his new wife on her household duties, tells his 15-year-old bride to cook some dish for as long as it takes to say a paternoster or a *misere*. Any frame of reference that would be understood by the listener would do: The Florentine chronicler Dino Compagni, writing between 1310 and 1312, describes a vermillion cross miraculously hanging in the air for the time it takes a horse to run twice around the piazza—an allusion, perhaps political, that while mystifying today would have been well understood by a contemporary.[21]

Interestingly, hourglasses, which we usually think of as simple and ancient devices to measure duration (so much so that they've become the emblem of the passing of time) made their appearance only in the fourteenth century. They seem to have first been used on ships because there someone is always on watch to reset the hourglass. Perhaps the earliest visual evidence of the hourglass is the figure of Temperance in the famous fresco *Good Government* in the Public Palace in Siena. The fresco was originally painted in 1337 but the hourglass was possibly added in the repainting of 1355.[22] Hourglasses were used in schools to time lectures by the end of the fourteenth century, as an illustration in a Hebrew Bible from Coburg from about 1395 shows.[23] However, if we think about it, the late adoption of hourglasses makes sense in a medieval context: they need constant attention, can measure only one short set period of time, and also need to be first calibrated against a known period of time. And, unlike church bells, they are *personal* markers of time and not *social* coordinators. This is fine if you're in a closed environment such as a ship but less useful in an urban setting.

To understand medieval timekeeping thus requires a recalibration of thought. Rather than *telling time* as we think of it today, for most medieval people, it was more useful to look to event markers or to be able to compare floating durations. Just as the length of daylight and the seasonal

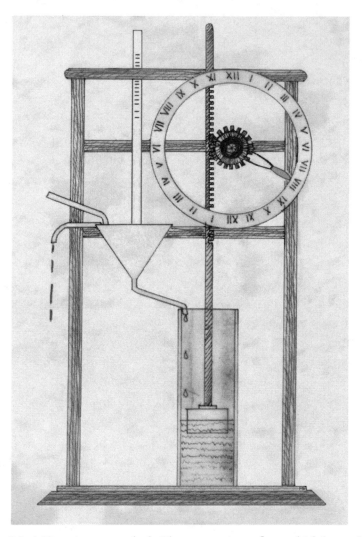

Figure 2.2. A Vitruvian water clock. The water raises a float, which has a shaft attached. The teeth on the shaft interact with a gear, which turns the hand of a clock-face. Wikimedia Commons

hours varied between summer and winter, so, too, did people's work schedules follow natural rhythms. Besides, prayer times floated throughout the year as the periods of light and dark lengthened and shrank. This is why, at least for purposes of social coordination, a timing device such as a water clock was more useful than a clock that tells the fixed hours.

Development of Water Clocks

A water clock is a sophisticated device for comparing durations that eventually gave rise to the mechanical clock. Also called a *clepsydra* (Greek for "water thief"), it may be more or less complex, but all water clocks are regulated by the flow of water into a basin. In the ninth book of his work on architecture, the same one in which he discusses his sundial, Vitruvius describes a float placed on top of the water basin that regulates the descent of a weight, which in turn drives a mechanical device such as a clockface or bells. (In figure 2.2, the water raises a float, which has a shaft attached. The teeth on the shaft interact with a gear, which turns the hand of a clock.) Vitruvius's work was recopied by the scholars employed by Charlemagne and his descendants in the eighth and ninth centuries and was not uncommon in medieval libraries, and so knowledge of this simple water clock was probably common.[24] Using a water clock to find the equinoxes was also mentioned by numerous late classical and early medieval writers.[25]

As with other timekeeping devices, the Arabic-speaking world had much more sophisticated water-clock construction than did the medieval West. Muslim savants were capable of creating rather complex water clocks with bells, alarms, and automata run by *epicyclical gearing* (gears mounted so that the center of one travels around the center of another), such as had been used in the Antikythera Mechanism. Some were also powered by streams and used waterwheels as a continuous source of power.

Another element these water clocks commonly possessed was an *escapement*, a device to control the flow of energy into the timekeeping mechanism. The importance of this invention cannot be overstated: without escapements, accurate timekeeping is impossible. In water clocks, this was a device to regulate the flow of water. It was not a new invention: for instance, in an escapement mentioned by the third-century BCE Greek writer Philo of Byzantium, the water from the reservoir drips into a small receptacle, which is balanced with a counterweight. When the receptacle is filled, it tips over and is reset. Such a device could be made to drive a clockface and would be marginally more accurate than a simple drip clock.

What Happened to Arabic Timekeeping?

Given the Muslim world's inheritance from the ancient world, aptitude for creating complex water clocks, and scholarly skill with both astronomy and astrolabes, we might think that the mechanical clock would have come about as a result of Arabic science. That it did not is mostly a result of cultural factors.

Historians of Arabic science note that there were really two sorts of astronomy in the Muslim world—folk astronomy, which depended on what the common person could readily observe in the sky, and scientific astronomy, which depended on systematic and carefully recorded observations. In keeping with the egalitarian nature of Islam, which does not favor placing critical religious knowledge such as prayer times in the hands of elites, scholars and judges favored this folk astronomy over the scientific. The Quran, after all, tells believers to look to the heavens and that the phenomena observed therein are given as signs for anyone who wants to understand.

Still, medieval Islam was tolerant and diverse, and even if it did not privilege scientific observations of the sky and, thus, astronomical timekeeping, it also didn't persecute those who engaged in such activities. Thus, it was not only medieval Christianity's linking of astronomical-chronometric knowledge and virtue but also its restriction of authority to a privileged few that favored the growth of scientific timekeeping in the West.

The Muslim calendar gives us an example of this tendency. As we saw in the first chapter, Islamic law uses a lunar calendar, with the beginnings

Conscious of their Hellenistic roots, Arabic sources often attributed the invention of such devices to Archimedes, but the peoples of the Near East were also innovators who used mechanisms such as escapements and epicyclical gearing to create massive public clocks, such as the one at the Umayyad Mosque in Damascus. Almost certainly a water clock, it was installed in the early 900s (CE) but had ceased working by the mid-twelfth century.[26] Other water clocks included a fourteenth-century example at Fez in Morocco, as well as those described in manuscripts by Muslim engineers and scientists. For instance, the Persian engineer Ismail al-Jazari (1136–1206), documents sophisticated clocks in his *Book of Ingenious Devices*, such as one shaped like an elephant that kept the

of months defined by the first sighting of the crescent moon. Because the moon had to be actually *seen*, the date could vary from place to place, depending on local conditions such as latitude and weather. Of course, scholars knew when the month *ought* to start, but this was a long way from actually regulating society according to scientific rules. Today, with no central authority to set a centralized calendar, Ramadan might begin on different days in different places. And, as I noted in the sidebar in the previous chapter, Mohammad forbade computations such as intercalary months, which relied on experts' specialized understanding of both the calendar and astronomical observations.

Prayer times in Islam were similarly determined by conventional signs, beginning with the evening prayer at twilight, followed by the prayer at nightfall, dawn, noon (after the sun has crossed the meridian), and afternoon, when an object's shadow has grown longer than what it was at midday by its own length. Those who called the faithful to prayer were chosen more for having a loud voice than for timekeeping skills. However, beginning in the thirteenth century, precise tables of the times for prayer began to appear, which required the use of astronomical instruments such as astrolabes, quadrants, and sundials to observe the sun, stars, and the length of shadows. Arabic astronomy—and, thus, timekeeping—retained its high degree of sophistication through the Middle Ages. However, it did not develop in the same direction as Western timekeeping because there was no perceived social need: accuracy and precision were not

(continued)

unequal hours and one shaped like a castle that kept track of astronomical phenomena.[27]

The ability to create elaborate mechanical clocks was another example of how the Muslim world was very much not only heir to, but also the continuer of, Hellenistic science and how the West is in turn indebted to Arabic science. In fact, Europeans seem to have been quite impressed by these devices. In 807, Harun al-Rashid, the ruler of the Umayyad Empire, sent Charlemagne a brass water clock from Baghdad that would sound bells and cause 12 figures of horsemen to move in and out of mechanical doors. The otherwise rather laconic authors of the *Royal Frankish Annals* describe this wonder in great detail. As another example, German chronicles speak of an immense mechanical

privileged over other ways of knowing, and observations were not seen as greatly important for social coordination. Also unlike in the West, we do not see an emphasis on timekeeping in the sense of coordinating objective measuring devices to the daily movement of the sun and the stars in the Muslim world, though tables for doing so existed. Finally, the West saw a vernacularization and spread of this knowledge, at least among those of some education. For instance, by the late Middle Ages, the Arabic technique of using shadow measurements to tell time had diffused as far as England, as we can see from Chaucer's observation at the beginning of "The Parson's Prologue":

> By the time the Manciple his tale had ended,
> The sun to the south line was descended,
> So low that it had not, to my sight,
> More than twenty degrees in height.
> Four o'clock it was though, I guessed,
> For eleven feet, or more or less,
> My shadow was at that time, as there,
> Of such feet as my length divided were
> In six feet in equal proportion.[1]

1. Larry Dean Benson, *The Riverside Chaucer*, 3rd ed. (Oxford: Oxford University Press, 1987), 287.

astronomical simulation given to the Holy Roman Emperor Frederick II in 1232 by Sultan al-Ashraf of Damascus; this is also noted in a trilingual inscription on the wall of Frederick's palace in Palermo, Sicily.[28]

It is clear from all this that Western scholars, and the rulers who patronized them, had a great hunger for Arabic scientific knowledge and technical know-how. Of course, mechanical clocks were not developed in the spirit of what we would recognize as modern scientific inquiry: the correct timing of religious rituals and the proper ordering of the world that these would have allowed were part of the duty of a saintly ruler.

Improved Water Clocks in the West

By the eleventh century, Western water clocks for keeping the unequal hours (the amount of water used varying with the season) were imitating the complexity, if not the splendor, of those in the Muslim world. To cite one example of a sophisticated clock, the biography of St. Wilhelm, who was abbot of Hirsau from 1069 to his death in 1091, describes a wonderful astronomical clock: "He left us many monuments of his natural genius, for he devised a natural clock from the example of the heavens. He showed with it the solstices and the state of the world by certain experiments. All his associates wrote letters to ask for it."[29] Similarly, in his *De anima* (*On the Soul*), written about 1240, Guillaume of Auvergne, bishop of Paris, recounts astronomers' use of water clocks that moved "by water and weights."[30]

Like their Islamic predecessors, these fancy water clocks often had bells or signaling devices attached to them. For instance, the biography of Peter Monoculo, eighth abbot of Clairvaux (1120–1186), has its hero first beset in a dream by a demon, whom he dispels thanks to the advice given by his mentor, Baldwin, and then awakened by the sound of the clock calling him to prayer, whereupon he prayed all night for Baldwin's soul.[31] To ensure that the call to prayer took place at the proper time, numerous monastic regulations specified that these clocks be carefully maintained, either by the sacristan (caretaker) or his helper. This nocturnal checking of clocks sets the scene for a sort of spiritual ghost story from thirteenth-century Italy: in the biography of Umiliana de' Cerchi (1219–1246), a pious woman from Florence who was attached to the Franciscan order and was later canonized, a monk in a Florentine monastery wakes up in the middle of the night two years after the saint's death to check the clocks (in the plural—*horologia*) to see whether it was time for matins and comes upon her ghost still hard at prayer.[32]

To be sure, precise timekeeping in a sense familiar to us today could not be accomplished with twelfth- or thirteenth-century technology. Bishop Guillaume mentions that the water clocks of his day required frequent repair. The water could freeze, wooden parts would crack and wear out quite quickly, and the medieval iron industry produced products of irregular quality. On top of that, water would flow out at different rates at different temperature because of changing viscosity. From a medieval context though, it is immediately clear that the use of a water clock is distinct from astronomical observations of the equal hour:

whereas observing the night sky and comparing observation against a known value can tell the observer the time in an absolute sense, a water clock, like repeating a well-known prayer, measures *duration*. The water, measured out according to the season, enters the basin, it fills up, the whole is reset, and one complete cycle is passed.

Again, we need to make a twofold division between *timing devices*—that is, those that can describe variable durations—and *devices that tell time*—that is, that mirror the motion of the stars, independent of any other variables. The water clock began as the former, but attempts to make it conform to the latter increased over the course of the thirteenth century. This drive toward greater accuracy, precision, and ease of use eventually gave rise to the mechanical clock.

The Invention of the Mechanical Clock

Much like the perpetual-motion machine that appeared in contemporary treatises, the *idea* of a reliable, precise, accurate, and convenient device by which one can know the motion of the heavens either day or night and, presumably, thus synchronize the functioning of human society and sacred prayer, was clearly present in Europe by the late thirteenth century. This would be a true clock—a device to mirror the motion of the outermost heavenly sphere, which was imagined to be the most regular, the most perfect, and the closest to God—in other words, to tell time in an objective sense. For instance, Robertus Anglicus, a professor at Montpellier, notes in his 1271 commentary on Johannes de Sacrobosco's c. 1230 *Treatise on the Sphere* (*De sphera*), the standard textbook on Ptolemaic thought known to all university students, that a good, accurate clock would keep time with the heavens:

> It may be noted that there are two kinds of hours in astronomy, equal and unequal. An equal hour is the twenty-fourth part of a natural day; wherefore, if an entire natural day were divided into 24 equal parts, then each of those parts would be called an "equal hour." . . . [A]nd we speak of the other hours, namely, the unequal, in the reckoning of hours by astronomical instruments and also by astronomical clocks [that is, water clocks that

are timing devices]. Nor is it possible for any clock to follow the judgment of astronomy with complete accuracy. Yet clockmakers are trying to make a wheel that will make one complete revolution for every one of the equinoctial circle [in other words, one revolution of the heavenly sphere, used for computing the absolute hour], but they cannot quite perfect their work. But if they could, it would be a really accurate clock and worth more than astrolabes or other astronomical instruments for reckoning the hours, if one knew how to do this according to the method aforesaid.[33]

Robertus Anglicus's commentary reveals several things. First, it tells us that, though the mechanical clock that kept equal hours was beyond the reach of medieval technology, the *idea* was there—much as science-fiction writers imagined space travel decades before it was feasible. Second, he imagines this clock, like the Alfonsine clepsydra, as a wheel traveling in time with the heavenly sphere. Third, it tells us that this device had not yet been invented by the 1270s. Fourth, it tells us that there was a sort of social demand for the equal hour, as also seen by the increased use of astrolabes. Finally, the mechanical clock was more valuable than a timing device such as water clock, because one could always reconcile the unequal hours to the equal with a simple table.

A contemporary example of a proposed device that could "make one complete revolution for every one of the equinoctial circle" can be found in the *Books of Astronomical Knowledge* (*Libros del saber de astronomia*) that King Alfonso X "the Wise" of Castile and Leon (r. 1252–1284)[34] had compiled in about 1276. (Alfonso also had Ptolemy's *Almagest* newly translated from the Arabic and a new set of astronomical charts called the *Alphonsine Tables* drawn up.) The *Books of Astronomical Knowledge* contain several designs for clocks, including a treatise written by a Rabbi Isaac ibn Cid that details the design for one—actually, a cylindrical clepsydra—in which mercury percolates between 12 chambers arranged around the rim of a large wheel, thus controlling the descent of a weight. The device, which resembles both the waterwheels known throughout the Muslim world and a design discussed by Ptolemy and elaborated upon by the English Franciscan friar Roger Bacon

The Invention of the Mechanical Clock in China?

It is worth taking an aside here to comment on the possibility that the first mechanical clocks came from China. This idea was introduced in the 1940s and 1950s by the British biochemist and historian of science Joseph Needham (1900–1995), who wrote that the first records of Chinese water clocks date from the sixth century BCE, that the philosopher Huan Tan (43 BCE–28 CE) observed the discrepancies between sundials and water clocks due to humidity and temperature, and how other Chinese savants experimented with mercury-driven and sand-driven clepsydrae. Needham also discovered records of several Chinese astronomical clocks from the tenth and eleventh centuries.[1] Chinese astronomers, he argued, were no less inventive than European astronomers, and they further benefitted from the fact that—unlike the West after the decline of Rome—there was no widespread civilizational disruption in China. As we saw in the previous chapter, the Chinese were also sophisticated astronomers and timekeepers.

Yet modern technology did not emerge in China—or, as Needham puts it in what has been called his "Grand Question," "Why did modern science, the mathematization of hypotheses about Nature, with all its implications for advanced technology, take its meteoric rise only in the West at the time of Galileo?"[2] This question is closely tied with the history of timekeeping, the primary technology based on the primary

(1220–1292), is designed to turn a star map constructed like the rete of an astrolabe by means of a reduction gear calibrated so as to be in time with the revolution of the heavens.[35] While it is not known whether the Alfonsine clepsydra was ever built, the *idea* of a continuous drive that operates by regulating a falling weight was there, and the historian of science Derek Price speculates that this device inspired the weight-driven mechanical clock. The weight-driven mechanical clock might have been an attempt to make a cheaper alternative to a mercury clock: the only place in Europe where metallic cinnabar (mercury ore) was known to exist in abundance was the Iberian Peninsula. Certainly, this technically intriguing idea of timekeeping seemed plausible, and a water-driven version would reappear in the late sixteenth century in a short treatise by the Italian writer Attilio Pariso.

science, astronomy. Needham himself came up with several explanations for his meta-question, as have the historians who have succeeded him.[3]

To understand Needham's arguments about Chinese science, we need to understand something about ancient Chinese philosophy and political theory. Much as in the West, the purpose of Chinese astronomy (and, thus, timekeeping and calendars) was to demonstrate the interrelation of the human, natural, and divine realms. It was therefore important to correlate what was observed in the heavens to what was known on earth. For instance, the year had 365 days; so, too, did the human body have 365 joints. This sort of syllogistic reasoning was all-pervasive in classical Chinese thought, as we can see from works such as the *I Ching*, where the hexagrams have correspondences with natural elements and numbers. Similarly, concepts such as colors and music are also associated with number. Finally, the calendar had a ritual purpose: to keep the Mandate of Heaven, the divine right to rule, Confucian rituals had to be performed at the proper time. Issuing a calendar was both the prerogative of, and a propaganda statement by, the imperial court.

This was, of course, not unlike the linking of religion, astronomy, and timekeeping in the Western tradition. The difference is that, in China, the calendar was issued by an official, centralized bureaucracy, whereas the fractured West, though nominally unified by the church, had both the practice of making independent observations and a lively

(*continued*)

But the invention of such a machine would have implications that were not yet realized. The knowledge it would create had the potential to be abstracted from any external referent, such as the movement of the stars. The equal hours marched on, unchanging; theoretically, the device was showing this abstract knowledge instead of modeling the physical universe. Rather than mirroring the movement of the heavens, a clock produced a count that was entirely self-referential, divorced from any observable event. Time was on its way from being a dependent variable, a measure of the duration of phenomena, and becoming an independent variable abstracted from any material object.

The mechanical clock was invented in about 1300 or shortly thereafter. A more certain date cannot be given, because of a lack of precise sources: vague references such as Dante's verses, a mention of an tradition of debate inherited from the Greeks. True, there was an inarguable framework—the date of Easter, for instance, would always be a constant, at risk of heresy—but each observer also had the ability to determine the equinox on their own.

An example of this difference in mentality is shown by the embassy Confucian scholar Su Song made to the Khitan people north of China in the late eleventh century. Su Song was supposed to arrive on the Khitan emperor's birthday, which coincided with the winter solstice, but the Chinese calendar was running a day behind the Khitan. When the Khitans pointed out the embarrassing error, Su Song, in his defense, could only hold up the calendar promulgated by the official Chinese court astronomers. But keeping objective time—knowing the mathematical solstice—wasn't the priority for the Chinese court; showing the proper time for ritual observations was. As the economic historian David Landes puts it, "Knowledge of the right time and season was power, for it was this knowledge that governed both the acts of everyday life and decisions of state. . . . [T]his was a reserved and secret domain." Landes concludes that this "recourse to authority rather than to evidence" led to "the corruption of science by factional politics."[4]

The greatest development of Chinese astronomy was the "clock" that the presumably chastened Su Song constructed in the late eleventh century. Though Needham promoted the idea that this Chinese invention might have been the original clock that inspired European timekeeping technology, this is unlikely. Besides the fact that the West had ample examples from the Muslim world, Su Song's device was too difficult to maintain; the rulers and bureaucracies of later dynasties lacked sufficient interest in precise timekeeping; and, most importantly, the Chinese "clock" was not a clock as we are familiar with it but an immense water-powered model of the heavens (an *armillary sphere*), some 40 feet high, which replicated the motion of the heavenly bodies. To be sure, it was brilliantly constructed, with gearing and multiple water reservoirs acting as escapements to make it as accurate as possible. However, its very complexity made it difficult to maintain and impos-

apparently all-mechanical clock made of silver that supposedly belonged to King Philip the Fair of France (d. 1314) in a 1380 inventory of his successor Charles V, a poem by the mystic Heinrich Suso called the "Clock of Wisdom" ("Horologium sapentiae"), and ambiguous references in monastic chronicles are all we have to go on.[36] The replace-

sible to reproduce. Besides, it had no lasting influence: Su Song's clock was damaged in 1127 by the Jin invaders who sacked Kaifeng and ended the Northern Song dynasty, then finally abandoned by the Jin in the confusion of the Mongol invasions.

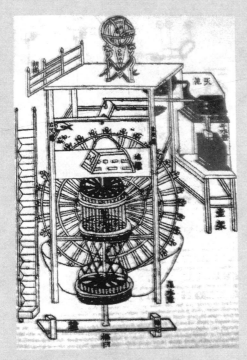

Figure 2.5. Su Song's clock.

In the end, Su Song's clock was a spectacular, singular device made for a centralized authority. China did not march down the same road Europe did because it did not have the same perceived needs: it was a stable, prosperous, and well-governed society with a calendric and

(*continued*)

ment for the clepsydra was a mechanism that became standard until the mid-seventeenth century. This was the *virge-and-foliot escapement*. (For the sake of not detracting from the narrative, I will explain the various clock mechanisms I mention in this book in detail in the glossary.)

daily timekeeping system that worked quite well, while the decentralized West was comparatively politically and intellectually disordered.[5]

This is not the only possible explanation for why history turned out the way they did. Other scholars have posited equally convincing arguments, and Nathan Sivin of the University of Pennsylvania has even asked whether the "Great Question," with its assumption that "progress" is a single straight line, deserves to be posed. China—or the Muslim world—wasn't "better" or "worse" or more or less "advanced" than the West—it was a different society whose different perceived needs took it down a different path.

1. Most notably, his *Science and Civilization in China* (Cambridge: Cambridge University Press, 1954–) and *Heavenly Clockwork: The Great Astronomical Clocks of Medieval China* (Cambridge: Cambridge University Press, 1986).

2. See Joseph Needham, *The Grand Titration: Science and Society in East and West* (Toronto: University of Toronto Press, 1969), 16. For a thorough historiographical overview of the question as it stood in the mid-1990s, see Hendrik Floris Cohen, *The Scientific Revolution: A Historiographical Inquiry* (Chicago: University of Chicago Press, 1994), 378–505. I am also drawing on David S. Landes's rebuttal of Needham in *Revolution in Time* (Cambridge, MA: Harvard University Press, 1983); and Geoffrey Ernest Richard Lloyd's treatment of Chinese historiography in *Ambitions of Curiosity* (Cambridge: Cambridge University Press, 2002). For another important perspective, see Nathan Sivin, "Why the Scientific Revolution Did Not Take Place in China—or Didn't It?," *Chinese Science* 5 (1982): 45–66, revised August 24, 2005, http://web.nchu.edu.tw/pweb/users/hbhsu/lesson/8252 .pdf, accessed July 5, 2017.

3. Cohen, *Scientific Revolution*, 439–466.

4. Landes, *Revolution in Time*, 33.

5. One variation of this argument is made by Justin Lin, "The Needham Puzzle: Why the Industrial Revolution Did Not Originate in China," *Economic Development and Cultural Change* 43, no. 2 (January 1995): 269–292. See also Etienne Balazs, *La bureaucratie celeste: Recherches sur l'économie et la société de la Chine traditionnelle* (Paris: Gallimard, 1988). Balazs blames centralized, stifling state control for choking innovation. I have personally been influenced by Jared Diamond's arguments in *Guns, Germs, and Steel* (New York: W. W. Norton, 1997): China's geography made early and continual unification practical, and periods of disorder were comparatively few; complimentary mentalities led to China being, in Needham's words, "homeostatic" but not "stagnant."

An interesting precedent to the virge and foliot was the *strob escapement*, such as the one famously designed by Richard of Wallingford, abbot of St. Albans in England. Richard was born the son of a blacksmith in 1292, orphaned and adopted by the abbot of St. Albans at the age of 10, graduated from Oxford in 1314, became the 28th abbot in 1327, and died of leprosy in 1335. In his brief tenure, he not only reorganized and renovated the abbey but also composed works on mathematics. His prototype clock does not survive, but several modern reconstructions show that it was a workable design that probably would have rung twenty-four-hour time. Interestingly, Richard's other passion was trigonometry. His mathematical studies would have been invaluable in constructing the gearing for his clock.[37]

Adoption of the Mechanical Clock

Richard of Wallingford was hardly unique in his efforts, as mechanical clocks were springing up all over Europe during his lifetime. In 1322, a clock was constructed in the Church of St. Catherine in Rouen in northern France that could be easily heard in Roncherol, some five kilometers away. This was undoubtedly a mechanical clock, since it was made to play a hymn when it struck.[38] In the same year, a clock with an astronomical dial and automata was begun in Norwich Cathedral in England; the cathedral sacristan's roll gives us a detailed account of the expenses. Other early French mechanical clocks were located at Caen, Valenciennes, and Beauvais. In 1326 a mechanical clock with automata—a procession of those famous astronomers, the Three Wise Men—was built at the hospital of St. Jacques in Paris, a major lodging-point for the faithful on the way to the pilgrimage Church of St. James of Compostela.

Italy, with its large urban centers and skilled artisans, had a large number of early clocks. Certainly, Italian merchants were acutely conscious of time: even as early as the thirteenth century, notaries in Genoa tended to record the canonical hour in which a contract was drawn up.[39] We have references to an iron clock being installed at Basilica of Sant'Eustorgio in Milan in 1309 and a clock-keeper being employed in 1322 in Ragusa (modern Dubrovnik in Croatia, which was then under the rule of Venice). The first known public clock to strike the equal hours was Italian, constructed between 1330 and 1336 in the Church of St. Gottardo in Milan and sponsored by Azzone Visconti, lord of that city.[40] Some eight years later, Ubertinello, the ruler of Padua commissioned the

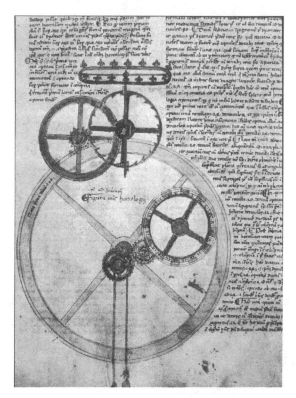

Figure 2.3. The Dondi Astrarium.

physician, astronomer, and natural philosopher Jacobo Dondi to make a similar public clock. Dondi's son Giovanni shared his interest in clock building, and his detailed description of the astrarium, an astronomical clock, survives in 12 manuscripts. As the name indicates, the purpose of the astrarium was not mere time telling but constructing a model of the heavens. Indicators for astronomical phenomena were common in clocks all through the medieval and early modern periods.

This is not to say that medieval timekeeping was anything like modern timekeeping! In fact, there were many local conventions for marking the hours. The system of "Italian hours," which were kept not only in Italy but in some parts of Bohemia (the modern Czech Republic) and even Poland, reckoned the day in twenty-four hours that began at sunset. Seen from a modern perspective, the answer to the question, "What time is it?" would thus vary throughout the year. The Italian system lasted until the mid-eighteenth century and even beyond in certain places. (Conversely, our modern schema of beginning and ending the day at midnight was known as "French hours.") Medieval clocks lacked a minute hand—which would have been superfluous, as they were nowhere near accurate enough to keep such precise divisions of time. We should thus by no means confuse the desire to construct a device to mirror the heavens—that is, keep track of the sidereal day, and thus the equal hours—with modern timekeeping.

The mechanical clock, like other timekeeping devices, quickly acquired moral meaning and became tied up with the appearance of virtue and the right to rule. Dante Alighieri (1265–1321), in the 10th canto of his *Paradiso*, considers the heavens as the clock that calls one to Matins, and in the 24th he sees his beloved Beatrice dancing with the other blessed like the gears of a clock. A little later, the mystic Heinrich Suso gave the Latin translation of his *Little Book of Eternal Wisdom* (*Das Büchlein der ewigen Weisheit*, written between 1327 and 1334) the title of *Horologium sapientie*, or *Clock of Wisdom*. In the next generation, the chronicler Jean Froissart, in his poem "The Amorous Clock" ("L'horloge amoureuse"), written c. 1369, made the parts of the escapement clock into an allegory of love—incidentally giving us an excellent description of the inner workings of a medieval clock. The first gear, signifying desire in the heart of man, is moved by the weight of beauty, the virge and foliot govern all, the dial profitably shows the results, and the clockmaker can enjoy the ringing of the bells without putting the whole out of order. Time telling was a means of setting the world in order and something to which someone with a godly mind ought to turn his or her attention.

Similarly, in 1372, the Carmelite monk and theologian Jean Golein (1325–1403) translated Guillaume Durand's thirteenth-century *Explanation* into French. In his sumptuous manuscript, which survives in two versions, Golein adds in the middle of Durand's note that it was Pope Sabinian (604–606) who ordered that bells be sounded in churches on a twelve-hour system:

And King Charles ordered for the first time that this be done in Paris by "points" [quarter hours] by the clocks he had erected at his palace and in the woods and at St. Paul. And he had workers come from strange places at high prices to have this done, so that the clergy and other people could know the hours and have proper customs and devotion to serve God. That is to say, formerly they were sounded once at Prime and twice [or thrice] at Tierce, so that one did not have certain knowledge of the hours as one has [now], and one can say of this of Charles the Fifth, king of France, that *a wise man is ruled by the stars*, for one can know the hours without fail whether there it is sunny or not by these well-adjusted clocks.[41]

A hundred years after Durand's original, we see the change in mentality from the traditional floating canonical hours to "certain knowledge."[42] Ambiguity has been vanquished; rather than having to rely on variable and fallible human signs, we have at least the idea of an objective time-keeping device by which human life can be regulated. So long as there was a skilled artisan maintaining it, the mechanical clock was easy to use—one only had to listen for the ringing of its bells—and it was perceived to be more accurate and more precise than the human-centric measuring of durations with water clocks that had preceded it. It therefore acquired the authority traditionally given to those who watched for signs in the stars.

Regulating Urban Life by the Clock

Given this increased accuracy and precision, it unsurprising that mechanical clocks such as the one established by Charles V quickly became used to regulate urban life. For instance, a Parisian order dating from before 1386 required law courts to be open from "nine o'clock according to the palace clock or thereabouts" to noon between the feast-day of St. Remigius (October 1) and Easter, and from eight until eleven from Easter to October 1.[43] Compare this to a 1327 regulation that simply stated that court would begin "after sunrise or within the space of time thereafter not to exceed a short mass."[44] In Paris, as well as in other cities, clocks also became a tool for at least some shop owners to squeeze as much work as possible from their employees. A 1384 regulation of the *tondeurs de draps* (the cloth cutters' guild) demanded that wageworkers labor every day and every hour other than solemn feasts and those of the Apostles. From the Feast of St. Remigius to Candlemas (February 2), they were required to go to work from twelve o'clock at night until daylight, whereupon they had a break until nine o'clock. There was a further one-hour break at 1 PM, and then they worked to sundown. The rest of the year, they worked from sunup to nine o'clock in the morning, then had a one-hour break, and then worked to one o'clock in the afternoon, when they had either one or two hours, depending on the length of the workday, for eating and recreation. They were then required to return to work until sunset, at which time they had a half hour to drink and refresh themselves in their master's house. They were further commanded to not quarrel about the work times and to keep track of them daily.[45] (Note that, unlike our current system of changing the *time* for daylight savings, medieval people changed the appointed *hour*.)

University Time and the Philosophy of Time

The influence of medieval timekeeping technology, and the new methods of social coordination it engendered, is evident in what medieval philosophers wrote about the nature of time. In turn, these theories of time would affect later developments. Think of any graph you may have studied in a science class: the *x*-axis almost invariably depicts the timescale along which the phenomenon depicted on the *y*-axis takes place, be it a chemical reaction or the trajectory of a falling object. Time is, as we say, the "independent variable." The idea of time proceeding in an endless march of independent, identical units is, of course, a fiction, but it is a fiction that makes possible our understanding of physical world. The rise of modern science required both precise measurement and the mental constructions to go along with it, and the groundwork for both was laid by medieval scholars.

That the universities that such scholars worked in were subject to tight temporal regulations should be no surprise to modern students. Much as today, running a medieval university required a great deal of coordination. Scheduling faculty meetings, examinations, and lectures was as necessary then as now. In addition to these problems were added particularly medieval concerns such as holding common prayer times and debates. Moreover, regulations had to specify *which* signals from *which* bells should be heeded. Thus, by the nature of their daily routines, the members of the university tended to be more conscious of time—and at an earlier date—than other segments of society.

The medieval college curriculum was also inextricably tied up with the mental tools needed for timekeeping. The usual course of study consisted of the basic knowledge called the *trivium* and more advanced studies called the *quadrivium*. The trivium were the arts of language— grammar (that is, Latin), rhetoric (how to speak in public), and logic (how to make a good argument). The quadrivium can be understood as the study of number—arithmetic, which is pure number; geometry, which is number in space; music, which is number in time; and astronomy, which is number in space and time. All of these were immensely useful to a career in the church and the increasingly secular royal bureaucracies—the trivium for preaching and the quadrivium in fields as varied as architecture, singing mass, observing the times for prayer and holidays, and for understanding the astronomical basis of timekeeping or geometrical methods of constructing clock gearing.

From the twelfth century on, this curriculum was also based on the study of Aristotle—first in translation from Arabic to Latin and then in translation from the original Greek. Aristotle was considered the preeminent authority on the natural world—so much so that he was often known simply as "the Philosopher." His ideas, and the need to reconcile them with Christianity, gave medieval academics a lot to argue about, which was all to the good because the medieval university thrived on disputations. (In fact, to be promoted and graduate, you had to perform well at debates.)

One of Aristotle's major concerns in the *Physics* is the nature of time. Without motion, he says, there is no perception of time; however, time is not motion, because if we are in darkness, sensing nothing with our bodies, our minds or souls (*anima* in the Latin) still perceive time. If time is to be measured, then the smallest number is two, because we must have two quantities to compare to one another. Just as one can measure a line only by comparing it against another line, so, too, must one have two times to compare with one another. In other words, we measure longer times by shorter, just as we measure a herd of 100 horses as consisting of 100 individual, discreet horses. To Aristotle, then, time is the "number of the motion with respect to the before and after"—in one Latin translation read by medieval scholastics, "numerus motus secundum prius et posterius." The very essence of time is measuring it.

To appreciate how this worked "on the ground," we can see medieval philosophies of time reflected in a thirteenth-century regulation from Paris on how university officers should be elected:

> Item, that the election of the rector shall last the duration of one candle. This shall be a regulated candle without deception of one pound of wax above a candle-holder of a weight of eight new silver coins, made with four wicks, as is commonly used, so that those who use it can have it made without difficulty, divided in 26 parts, each of which shall be an eighth of a Parisian yard [ulna]. If a lesser weight of wax is used, it shall be of a similar proportion as above. This candle shall be lit in the entrance of the place where the election is held, and shall continue to burn until it is consumed. If, however, anyone takes away or adds to this, they shall be ineligible for any office . . . for the time of the rector's tenure.[46]

In other words, if someone were to ask, "How long was the election to last?" the answer would be, "one precisely constructed candle of such-and-such a weight"—*not* a quantity of hours or minutes. The consumption of the wax of the candle, observed by the members of the university, is the change—the movement—by which time is to be reckoned. Of course, university professors were not only ones to use candles, but the university regulation accords well with medieval theory on the nature of time: to medieval philosophers, time was measurable only by the comparison of durations by intelligent beings.

Another example of applied Aristotelianism was in fencing: in medieval German treatises on fighting with the longsword, fencing masters speak of the "before," "after," and "during" of an action. What this refers to is the relative timing of an action vis-à-vis the adversary's movement. The duration of each technique—whether long or short—depends on the adversary's own actions. One must perform the right action at the right time: if the opponent strikes and you don't parry in time, you will be hit; strike when he or she is off-balance, and you will gain the advantage. To emerge the victor, one must have a sense of timing, knowing both when to initiate an action and how large or small, fast or slow, the action should be. To teach this, one also needs a language to specify the configuration of actions in space and time. Unsurprisingly, the symbol of the fencing master remained the compass dividers—the symbol of proportional measurement—well into the Renaissance.

The Medieval Philosophy of Time Before and After the Mechanical Clock

The burning of a precise candle, or describing the "before" and "after" of sword-fighting techniques, illustrates how medieval commentators interpreted "the number of motion" in Aristotle's *Physics*. For medieval thinkers before the mechanical clock, time is something that we can measure only by proportionately dividing the movement of material objects. However, in the 1300s, after the invention of the mechanical clock, there was an increased emphasis on tying the true nature of time to the "outermost heavenly sphere" and an increased emphasis on trying to find an "objective" time and the equal hour. For instance, the English Franciscan William of Ockham (1287–1347), writes that "the motion that should be counted as time is that of the primum mobile, which has the most uniform and the swiftest motion, for though we can measure the motion of many things by the motion of the sun . . . prime motion is . . . better to measure other motions."[47] That is, the unequal hour, the time of the sun, needs to be corrected to the "true" sidereal time of equal hours. Time, in other words, is identical with the very thing that clocks were made to correspond with.

Similarly, the Parisian philosopher Jean Buridan (c. 1295–1363) recognized that time measurements in the human world are imperfect and inferior to those of the heavenly sphere. According to Buridan, "common people" tell time by the sun; when they cannot see the sun (common in northern France during the winter), laborers often use the duration of their work to tell the time, concluding from the amount of work finished that it is Tierce and time to eat; and also ecclesiastics use clocks to tell time—even if the movement of the clock is not time itself. However, clocks mirror the sphere of the stars, which, in turn, is driven by the primum mobile. The outermost sphere is not only the one that drives the motion of the rest of the universe but the movement from which we should derive our time cues.[48]

Buridan's student Nicole Oresme (c. 1323–1382) was not only a philosopher and academic but also an advisor to Charles V, who famously installed the first public clocks in Paris. He followed Buridan in his philosophy but also made some fascinating observations. For instance, it was Oresme who first expressed the idea of a "clockwork universe" in his book *On the Heavens* (1372–1377). In the same book, Oresme introduces the so-called traveler's dilemma: Suppose three priests live in a city. Nine days before Easter, one sets out eastward along a road that goes around the entire earth, another sets out westward along the same road, and the third stays put. Both travelers circumnavigate the globe and come back home on the day that the stay-at-home-priest celebrates Christ's resurrection. However, the priest traveling westward has counted 10 days, returning home before Easter, while the one traveling eastward has counted only 8 days. Clearly, they are all in the same "time"—the date of Easter is an absolute and it would be heretical to say otherwise—but one has apparently traveled forward and one backward in time! The movement of the heavenly sphere, according to Oresme, is therefore not an absolute chronograph. Rather, time is something abstracted from material phenomena—something the physical world is measured by but that is not measured by the physical world.[49]

There is a certain irony, though, because, despite their seeming move away from "duration" and toward "objective" time, Ockham, Buridan, and Oresme were also part of a philosophical movement saying time could not exist without moving physical objects. This is a somewhat difficult concept to comprehend but crucial for understanding the emergence of modern science. It is the dispute between whether, as Aristotle holds, we can know the nature of things only from observing material objects or, as his teacher Plato (427–347 BCE) explains with the metaphor of the cave in his *Republic*, reality ultimately exists independently of human understanding.

For those not familiar with the cave, Plato asks us to imagine that we are prisoners shackled to chairs in an underground cave and made to observe shadows cast by a fire behind us. We would think that these "shadows" are real things—birds, trees, cats, and so on. If we were to get loose, we would turn around and see the puppeteers in front of the fire and realize we had been duped. Then, Plato tells us, imagine if we were to leave the cave entirely. Then, in the bright sunlight, we could discern the true nature of reality—real birds, trees, cats, and other objects. The idea that we can know the nature of reality only through abstractions is of no small importance to the development of Western science. Think of mathematical laws in chemistry and physics—ideas that can be touched only with our minds. Nonetheless, they describe the behavior of real phenomena.

The controversy arises with the problem of "universals." Think of dogs: Is there an essential "dogginess," a Platonic Dog, if you will, that all dogs participate in? Likewise, is there such a thing as Platonic "whiteness" or, rather, only white *things*, from which we derive a mental category of "white"? The former idea, which holds that there is indeed a Platonic Dog or a Platonic White, is termed "realism"—that is, that white has an objective, real existence. The latter—that it exists only in particular objects that we perceive with our intellect, from which we abstract an idea of "dogs" or "white"—is the Aristotelian position, called "nominalism" from *nomina mentalia*—"mental names."

The dogmatic Christian answer to the problem of universals is realist: not only does Augustine, who was a Platonist, place the Ideas (as these "real" things are called) in the mind of God, but there has to be some Platonically real essence of Christ to transubstantiate the Communion wafer into his body and blood. However, Aristotle was a strict nominalist: to him, the idea of "white" or the category of "dog" exists only in human minds.

The universals debate is directly applicable to timekeeping, since it raises the question of whether time is real, proceeding at its own pace independent of any other object—an idea we call "absolute time"—or whether, as Aristotle writes, it is "relative," measurable only in reference to observable moving objects. Oresme, in his thought experiment with the circumnavigating priests, goes the furthest toward suggesting an idea of absolute time divorced from observable phenomena such as the heavens. Yet he retreats from this, because, out of philosophical necessity, he and other medieval philosophers were compelled to tie time to the observable world. The traveling priests might have observed one day more or less than their stay-at-home colleague, but if they had instead measured time by counting silently to themselves during their voyages, they would have reached the same count. Time, in other words, is only a measure of relative durations. The idea that we can have sure knowledge of time was, in turn, one of the hallmarks of the emergence of modern scientific thought.

Is there such a thing as "time" independent from the material world, or, in keeping with nominalist thought, can its passage be known only through the durations or lengths of specific, known phenomena? Medieval thinkers such as Oresme needed to reconcile the idea of an objective measurement of time that exists independently of any observer (such as in the travelers' dilemma) with the durational idea of time dependent on the proportional measurement of a universe of moving objects (such as the rotation of the outermost heavenly sphere). In other words, to these fourteenth-century philosophers, time could be only the relative measure of durations.

Yet, by the mid-fifteenth century, some 70 years after Oresme's death, the German theologian Nicholas of Cusa (1401–1464) wrote in his *On the Vision of God (De visione Dei)*:

> For the simple concept of a clock enfolds all temporal succession. Now, let it be that the clock is the concept. Then, although we hear the sounding of the sixth hour before that of the seventh, nevertheless the [sounding of the] seventh is heard only when the concept gives the command. . . . And when we hear the sounding of the sixth hour, it is true to say that six sounds at that moment because the master's concept so wills it.[50]

For Nicholas of Cusa, the clock is the very image of God's Platonic knowledge of eternity: although human beings can experience time only as a durational succession of moments, this is simply because of our mortal limitations. The real nature of time is something outside of human perception. The idea of the clock itself—a machine for keeping time independently from any natural object—thus helped to birth the idea of abstracted, absolute time.

Limitations of Medieval Timekeeping

How precise and accurate was medieval timekeeping? Machining precise gears was certainly not out of the range of medieval technology: it requires only a compass, some basic geometrical knowledge, the ability to construct a jig, and the patience to file or cut the gears to shape. However, though it was an improvement over a water clock in that it is simpler, more reliable, and requires less constant maintenance, the virge-and-foliot escapement is far from perfect. The accuracy of even the best virge-and-foliot clock is only on the order of minutes per day, and it is vulnerable to humidity and the expansion and contraction of the mechanism due to heat and cold (though of course less so a water clock).

Perhaps the biggest hurdle medieval clock builders had to overcome was material: the pallets were always pushing on the crown wheel so that friction caused both wear and error to build up, and eventually moving parts would wear out. The same qualities that make good armor—toughness and hardness—make a good escapement, but medieval technology was simply not capable of creating iron plates of suf-

Figure 2.4. The virge-and-foliot mechanism.

ficient size for large gears until the improvement of the blast furnace in the early fourteenth century. It is not coincidental that we find early mechanical clocks in areas such as northern France that were also centers of iron production. It was not until the mid- to late fourteenth century that these blast furnaces turned out iron that was of sufficient quality to make armor; similarly, we also begin seeing the rise of large public tower clocks at this time. This is also likely why the first known tower clock, the St. Gottardo clock, was made in Milan: the city not only had an impressive infrastructure of watermills and the associated craftsmen to construct their gearing but was an armor-making center with ready access to high-quality materials.

The real importance of the virge-and-foliot clock is that it can measure the equal hours. The virge-and-foliot mechanism is also eminently well suited for producing a public striking clock, since placing it in a high place would provide advantage (there was more space for the weights to drop) against a water clock (where one would have to either transport the water against gravity or construct a long and inefficient drive-train to ring the bells). This is not to say that it was an automatic machine: medieval energy storage was too primitive, and any device would have required a constant input of work, either by refilling water basins or resetting weights.

There were also a number of problems with keeping accurate time in the Middle Ages. The utility of twenty-four equal hours was limited in a world without electric lighting, where the length of day in summer and winter varied so widely, people were more sensitive to the natural rhythms and their bodies' need for sleep, and work had to cease at sundown. In such a world, keeping Italian hours, or canonical hours, makes far more sense. No matter where the sun is in the sky, one knows that vespers is always in the middle of the afternoon. This is why many medieval clocks, such as the fifteenth-century example on the choir screen of Chartres, were made to keep both equal and unequal hours, so that users could identify the traditional time markers for their activities. They were, in a sense, mechanical astrolabes.

Without a doubt, the idea of equal hours gained currency through the fourteenth century. We can see this in numerous contemporary writings, from Francesco di Marco Datini, the famous "merchant of Prato," using terms like "twenty-first hour,"[51] to a 1396 French-English conversation manual asking "what time is it?" and expecting an answer according to twenty-four-hour clock,[52] to the time indications in chronicles such as Froissart's account of the Hundred Years' War,[53] to the monastery and theological college of St. Martial in Avignon fixing the offices "according to modern hour-reckoning" in 1379,[54] and, of course, the "four o'clock" of Chaucer's "Parson's Prologue."

Of course, the canonical hours continued to be rung, if perhaps slightly more precisely, and the older ways of reckoning time clearly persisted. In the biography of St. Coleta (1381–1442), abbess of an order of Poor Clares in Ghent, we read that a band of marauders sought to sack the town, "but the sacristan of the convent, who had to sound the call to Matins around the middle of the night, woke up between the 9th and 10th hour [9 or 10 PM], and believing it to be the middle of the night, in fact sounded the said call to Matins." The marauders believed the sound to be an alarm and (perhaps because of the large group of armed townsmen) repented of their evil. Miraculously, time sped up and dawn came the same amount of time after the nuns' prayers as it would have had the bell been struck at the proper time. Even though the town had a public clock, the convent had its own bell system, and our writer was apparently used to thinking about numbered hours, the sacristan still had to sound the signal in the "middle of the night"—that is, midway between dusk and dawn, proportionately considered.[55]

Despite such remnants of thought, clocks produced an important and indelible effect on late medieval society. By the late fourteenth century, mechanical clocks controlled the bells in medieval towns. These machine-controlled bells were much simpler to understand than the earlier cacophony of parish and monastic signals and necessarily simplified the ringing—it is easier to count 12 chimes than 24, for instance. Time itself became experienced as a result of mechanical processes, as a regular sign given by a machine. These regular public bells arguably produced a change in time consciousness at a general level: a device for measuring abstract time began to be used to regulate both personal and public activities. This could not help but open new imaginative vistas.

THREE

Savants and Springs

What then is time? If no one asks of me, I know; if I wish to explain to him who asks, I know not.
—St. Augustine, *Confessions* 11.14.17

GALILEO GALILEI had a conundrum: He wanted to prove that heavier objects fall no faster than light ones, but how was he to measure how long it took for them to reach the ground? Contrary to what many people think today, Galileo (1564–1642) didn't perform his experiment by dropping two cannonballs off the Leaning Tower of Pisa—at least not initially, though he did do that later on as a public demonstration. Rather, he rolled balls down inclined planes and timed their descent. But how was he to make an objective measurement of this? Galileo was living in in the early seventeenth century, and timekeeping technology hadn't changed much since the days of Oresme. Virge-and-foliot clocks, the state of the art, were not remotely accurate enough to keep seconds.

Finding a solution to the problem would do much to advance Galileo's astronomical arguments. Almost a century earlier, a German Polish philosopher named Nicolaus Copernicus (1473–1543) had challenged the classical idea of the universe. The earth is not the center of creation, said Copernicus, but rather, it, as well as the other planets, orbit around the sun. Galileo was a passionate defender of Copernicus's system of the solar system, which was still not entirely accepted by the scholars of his day. This was not mere dogmatism but quite rational objection. For instance, Copernicus's model required the stars to be unthinkably distant and therefore inordinately huge (because it was not known that the stars' apparent size is due to atmospheric refraction), and no one could explain what could possibly compel a mass of rock and dirt the size of the earth move through space at the speed required to orbit the sun once a year. Previous thinkers believed that the sun, moon, and other planets were made of sublime matter and stayed up in the sky by virtue of their lightness, while heavier objects sink down to the center of the universe (which to say, the earth's center).

Galileo's great contribution was providing experimental and observational evidence for what Copernicus had suggested purely as a mathematical model. For instance, he observed the moons of Jupiter through his telescope and saw them as a model of the earth-moon system and the earth-sun system. (They could also, he noted, be used as a sort of clock because of their regular motion.) Galileo also observed sunspots and the shadows of mountains on the moon and argued that celestial bodies were made of the same sort of ordinary matter that the earth is. This was the background to his challenge to the Aristotelian belief that heavier objects fall faster than light ones. The implication is that, rather than the planets and stars being remaining in the sky because of their lightness and less sublime, heavier objects sinking to the earth's center, some other force maintained the structure of the universe. Key to this would be providing experimental evidence that heavy objects fall just as fast as light ones.

To solve his dilemma in measuring how long it took the balls to descend, Galileo used an ancient technology: a water clock employed as a timing device. He discussed the exact method he used in his 1632 *Dialogue concerning the Two Chief World Systems* (*Dialogo sopra i due massimi sistemi del mondo*):

> For the measurement of time, we employed a large vessel of water placed in an elevated position; to the bottom of this vessel was soldered a pipe of small diameter giving a thin jet of water, which we collected in a small glass during the time of each descent, whether for the whole length of the channel or for a part of its length. The water collected was weighed, and after each descent on a very accurate balance, the differences and ratios of these weights gave him the differences and ratios of the times. This

was done with such accuracy that although the operation was repeated many, many times, there was no appreciable discrepancy in the results.[1]

There's a certain paradox here: though Galileo's experiment was crucial to the birth of modern science, and though it was used to help explode the neat harmony of the medieval worldview, it was also rooted in the practice and philosophy of medieval timekeeping. For Galileo, time was the ratio of the changes of moving objects. As Aristotle said, we can know the passage of time only by observing how one phenomenon in the world changes as compared to another—that is, by their movement. Galileo, in other words, was comparing durations in the medieval manner, not "telling time" in the modern sense.

Galileo was, incidentally, not the first to perform this experiment: the Flemish mathematician and engineer Simon Stevin had published his own falling-objects comparison in 1586.[2] Nor was he the last: astronaut David Scott famously repeated Galileo's experiment during his Apollo 15 mission to the moon in 1971. Since cannonballs would have been prohibitively expensive to take into space, he instead dropped a geologic hammer and a falcon feather and, indeed, they struck the moon's surface at the same moment.

Nor was Galileo the first to describe using a timing device for an experiment: Nicholas of Cusa proposed using similar clepsydra to find the relative durations of patients' pulses in his *Simpleton on Experiments in Statics* (*Idiota de staticis experimentis*). However, there is no evidence that Nicholas actually carried this experiment out.[3] Galileo's innovation, and the reason why his experiment is so important, is that he provides quite possibly the first record we have of a scientist actually using a timing device to conduct an experiment to accurately measure the actual duration of a physical phenomenon and thus provide evidence for proving, or disproving, a hypothesis.

The rest of this chapter will explore the contributions of Galileo, Huygens, Hooke, and other innovators before turning to Newton, looking at what he wrote in greater depth and its implications for the development of science, thought, and society. To Newton, time is not relative, dependent on the observation of moving objects such as Galileo's water timers

What Was the Scientific Revolution?

Because this is a book written for readers who may not have previously studied history deeply, it's worth taking a moment to explain what exactly the Scientific Revolution was. To begin with, it was far more than Copernicus's postulating the heliocentric theory of the universe and everything else, up to Newton's laws of physics, naturally following. Just as medieval thought was rooted in a particular culture and society, so, too, was the birth of modern science.

First, we need to recognize that the Scientific Revolution was firmly grounded in medieval thought. The quadrivium was the study of arithmetic (number), geometry (number in space), music (number in time), and, of course, astronomy (number in space in time), all of which had obvious implications for the mathematization of reality. Philosophically, Aristotle had insisted that we can only know things through the experiences of our own senses, while the Platonism that came back into fashion in fifteenth-century Italy insisted that true knowledge of reality was transcendental, equated with the Platonic forms that St. Augustine had placed in the mind of God. Medieval thinkers, following Euclid, believed that transcendental truth—the reality behind reality—could be shown through mathematics.

Experiment and the use of numbers to describe physical phenomena are the roots of the Scientific Revolution, but they needed to be connected to technology—the means to bring effects into being in the real world. Experimentation, such as Galileo's pendulum experiment, was key to the growth of science, but it required practical know-how that scholars didn't necessarily have—to build apparatuses, to grind

by an intelligent observer. Rather, time is "absolute," existing on its own, unchanging and evenly flowing. Newton's idea is the mentality of modernity, and for it to come about or be used in a practical manner required the development of more accurate, precise, and reliable timekeeping devices by the talented scholars and inventors who preceded him.

Galileo's Pendulum

Though Galileo used a medieval comparison of durations in his own experiments, it was another one of his discoveries that would eventually

lenses for telescopes, and do whatever else needed to be done. In fact, a thesis advanced by the historian Edgar Zilsel in the 1940s holds that the Scientific Revolution was birthed from the collaboration between academically trained natural philosophers and skilled craftsmen.[1] This is especially true for clocks, which require precise manufacturing techniques to construct.

Politics also played a role. Some of the controversy regarding Copernicus's heliocentric theory was attributable to the religious context: Martin Luther had published his 95 theses in 1517, and, ever since, Europe had been embroiled in conflict between those who wanted to maintain Rome's hegemony and those who wanted to break away. Galileo's well-known persecution came about not because he advocated the Copernican schema of the universe (which, his opponents argued, violated biblical descriptions) but because he engaged in amateur explanations of Scripture: "I think that in discussions of physical problems we ought to begin not from the authority of scriptural passages but from sense experiences and necessary demonstrations," as he wrote to the Grand Duchess Christina of Tuscany in 1615.[2] This was, of course, not acceptable to the Catholic Counter-Reformation.

We thus shouldn't be surprised that much of the process of discovery came about as a side effect of the Protestant Reformation. Luther himself wasn't necessarily pro-science—he is actually recorded as having made some disparaging statements regarding Copernicus—but he wasn't necessarily anti-science, either. More importantly, the zeitgeist of the time was one of questioning orthodoxy: If the authority of Rome could be overthrown, why not that of Aristotle? Finally,

(continued)

give rise to more accurate clocks—albeit not in his lifetime. This came about from his work with pendulums. Galileo found that, just as falling bodies accelerate uniformly, the period of a pendulum—the amount of time it takes to complete a swing—was, as far as he could tell with a small pendulum, independent of either the mass of the weight at the end (the "bob") or the degrees of arc of the swing. Galileo also found that the period of a pendulum varied according to the length of the string (I give the exact formula in the sidebar). However, as we will see below, Galileo was not quite right about the degrees of arc.[4]

Protestant theology said that works in this world are no less important than the spiritual life—an ideology that elevated investigating the
natural world and making fine instruments such as clocks into a sort
of religious calling. Protestant countries thus tended to be more
hospitable to science than Catholic ones—though we must not buy
into the myth that the Church was entirely hostile to science.

The displacement of traditional ideas didn't happen all at once,
either. Rather, it occurred in what historians now recognize as two
distinct phases. The first was the rediscovery of ancient knowledge
and of questioning the inherited system of the world. In this phase,
which ended with Galileo, scientists accepted that the truths of the
universe could be represented mathematically and learned to place
their faith in observation and experimentation.

The second phase of the Scientific Revolution, which built upon the
accomplishments of the first, centered on London in the late seventeenth
century. This was something of a golden age: the theocratic Common-
wealth that Oliver Cromwell had established to rule England after its
great Civil War (1642–1651) was overthrown in 1660, and, with the
restoration of the monarchy, the considerably more open Charles II came
to the throne. In the same year as Charles's ascension, the Royal Society
was founded under the king's patronage as a sort of "philosophical
club," with early members including luminaries such as Robert Hooke,
William Harvey, Robert Boyle, Christopher Wren, and, of course, Isaac
Newton. These men made enormous contributions to the disciplines we

The property of an event taking the same amount of time, independent of variables, is called *isochronism*, and it is why pendulums are so
valuable for timekeeping. Unlike the mechanism of a virge-and-foliot
clock, a pendulum will never speed up or slow down. In fact, in 1641,
the year before his death, the aged, blind Galileo came up with a design
for a pendulum clock, which he gave to his son Vincenzo to complete—
though Vincenzo was unable to do so before his own death in 1649.

Giovanni Battista Riccioli, an Italian Jesuit who taught at Parma,
went to rather extreme measures to verify pendulums' isochronism.
(The Jesuits, or Society of Jesus, is a priestly order founded by Ignatius
of Loyola in 1534 for the express purpose of combatting the Protestant
Reformation and spreading the Catholic faith. Their scholarly orienta-

now think of as biology, anatomy, engineering, chemistry, and physics. It was Newton's 1687 publication of *Mathematical Principles of Natural Philosophy* (*Philosophae Naturalis Principia Mathematica*) that marked the acme of the Scientific Revolution's achievement and the culmination of the second phase of the Scientific Revolution.

The overall significance of the *Principia* is this: Newton explains in mathematical terms the workings of the solar system as observed by astronomers such as Tycho Brahe and Johannes Kepler. The power of the human mind, in other words, could explain and even order God's creation. This established a paradigm—human reason as the ordering force in the universe—that would last for more than 300 years. These changes were reflected in politics, as well: in 1688, James II, brother and successor to Charles II, was deposed in the Glorious Revolution, firmly establishing Parliament's supremacy over the monarchy. The following year, John Locke published his *Two Treatises on Government*, giving an intellectual framework for nascent democracy. Now, in civic life, as well as scientific life, reason would rule over tradition. We call this intellectual movement that displaced religion with reason, tradition with science, and medieval kingdoms with democratic nation-states the *Enlightenment*, and it is the basis of our modern world.

The Enlightenment ultimately took its justification from developments in natural science. As I discussed previously, the Scientific Revolution was rooted in astronomy, with the initial step being

(continued)

tion made them not only ideal missionaries but also educators and researchers; accordingly, they figure large in the Scientific Revolution.) In his 1651 book *The New Almagest* (*Almagestum Novum*) Riccioli reports using the transit times of stars to measure the swinging of the pendulum.[5] His goal was to produce a pendulum with a swing time of precisely one second, which would complete 86,400 swings in a day. He ensured its accuracy by recruiting nine of his fellow Jesuits to watch his pendulums over the course of a day and count the swings. Of the nine, seven quit because of exhaustion. He then used his new timer to repeat Galileo's falling-bodies experiment, which he verified—an act of no small significance, considering that Riccioli was an anti-Copernican. Like many, he was unconvinced that Copernicus's defenders had sufficiently countered their detractors' arguments, especially regarding the

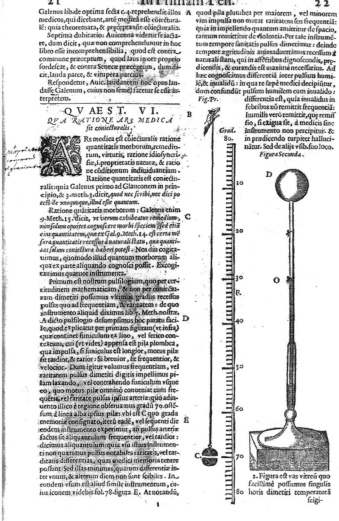

Figure 3.1. Santorio's pulsilogium (center), at the right is a thermometer.

necessary size of the stars, which he himself verified by astronomical observations. However, Riccioli was also willing to reject Aristotle where Aristotle conflicted with what he had observed.

Because of their isochronic property, pendulums quickly found their way into use as timing instruments even before the invention of an ac-

Copernicus publishing his heliocentric theory in 1543 and the capstone being Newton's *Principia*. Why the significance of astronomy? To begin with, astronomy, after all, is number in space and time. Plus, by observing the heavens, philosophers thought that they could read God's plan for the cosmos—and if astronomy were amenable to reason, why should human society not be, as well? Accurate astronomy, however, required more accurate clocks. Thus, timekeeping was critical to the birth of modern science and all that followed—including the ordering of human affairs by the clock. Similarly, clockwork suggested new machines to manufacture goods, the possibility of new means of organizing production, and all that followed. I will discuss timekeeping and the beginnings of industrialism in the next chapter.

It is also in the era of the Scientific Revolution that we see a continuation of our main theme: even as the technical means of measuring time grew more sophisticated and clockmakers and scientists continued to come up with new inventions and discoveries, there was a continued drive for simplicity, ease of use, and accuracy. Innovation, rather than being the product of lone geniuses thinking "ahead of their time," was driven by growth in technical capacity, such as improved metallurgy and tools; social factors, such as the increased desire to regulate society and the Protestant Reformation that overthrew the authority of the Church, which was one of the keystones of

(continued)

curate pendulum-based clock. For instance, the physician Santorio Santorio, a friend of Galileo's, described measuring patients' pulses with a "pulsilogium" in his 1631 book *Methods of Avoiding All Errors concerning the Medical Art* (*Methodi Vitandorum Errorum Omnium Qui in Arte Medica Contingent*).[6] The pulsilogium worked by synchronizing a pendulum to the patient's pulse by adjusting the string and measuring the length to get a relative idea of how fast or slow the pulse was. Thus, a period of time was converted to a measurable length—an ingenious solution in an era before stopwatches. The pulsilogium was in some ways also the first use of instrumentation to measure the human body—preceding X-rays, EKGs, MRIs, and other such staples of modern medicine (though Nicholas of Cusa was earlier, again, there is no evidence that he actually conducted this experiment, and the water clock was not a specially made

the medieval structure of the universe; and the questioning of received knowledge and drive for new understanding that was part and parcel of trying to achieve a new model of the universe.

What is the larger significance of this? Again, religious requirements for timekeeping led to precise observation of the sky. Even the Catholic Church, which we may think of as being "anti-science" for its persecution of Galileo, was forced to make precise observations of the natural world to properly observe its own festivals. Again, we see the self-reinforcing nature of developments in astronomy and time-keeping and the sixteenth century's demands for increasing precision. These developments were carried over into the social sphere: clock gearing suggested not only the inner workings of the machines that would be critical to the nascent Industrial Revolution but the possibility of ordering society itself like clockwork.

1. Edgar Zilsel, *The Social Origins of Modern Science*, edited by Diederick Raven, Wolfgang Krohn, and Robert S. Cohen (Dordrecht: Kluwer Academic, 2000); recently reinvigorated by Pamela O. Long in her *Artisan/Practitioners and the Rise of the New Sciences, 1400–1600* (Corvalis: University of Oregon Press, 2011).

2. Katharine J. Lualdi, *Sources of The Making of the West*, vol. 1, *To 1750: Peoples and Cultures* (Boston: Bedford/St. Martin's, 2012), 308.

Galileo's Pendulum Formula

According to Galileo, the formula for the period of a pendulum (t) is equal to 2π times the square root of the length of the string (l) divided by the gravitational constant (g)—or, expressed mathematically:

$$t = 2\pi \sqrt{l/g}$$

Since the only variable is the length of the string (unless you were to change the gravitational constant by moving the pendulum to another planet), this gives you the formula for the period.

You can perform Galileo's pendulum experiments for yourself with the chapter exercise at the end of the book.

apparatus). Some of Santorio's illustrations show a dial that could let out or take up the length of the string, thus altering the period.

Riccioli's and Santorio's work shows the importance of precise time-keeping to the Scientific Revolution, but, like Galileo, their work was also rooted in medieval thought: just as Aristotle held that time existed only when change is observed by an intelligent mind, so, too, were Riccioli

The Gregorian Calendar

The Gregorian calendar[1] reform is an example of how the journey to greater simplicity and precision doesn't follow a simple or precise narrative. In this case, we can see how the Catholic Church was open to innovation—quite against what is generally believed about the church being a reactionary force—while Protestant nations often weren't.

Because Easter is the holiest day of the Christian liturgical year, its proper observation was an important matter: celebrating the holiday on the wrong date was tantamount to heresy. For centuries, Julius Caesar's calendar, together with a special lunar calendar, had guided the yearly cycle of Christian liturgy, but by the late sixteenth century it was no longer considered accurate enough.[2] Because the 19-year Metonic cycle was off by one day every 310 years—nothing that would be noticed in one human lifetime—there was a four-day discrepancy between the observed moon and the calendric lunar cycle by the late 1500s. Since March 21 was considered to be the equinox for purposes of the church calendar, this meant that the actual Easter celebration was off quite a bit from what it was supposed to be.

There was a great deal of debate over how to correct the calendar. The solution was a compromise proposed by the physician and astronomer Aloysius Lilius (c. 1510–1576) in the 1570s, with the great Jesuit astronomer Christopher Clavius (1537–1612) being chosen to refine it and put it into practice. Pope Gregory VIII ordered the new calendar adopted on February 24, 1582, with his bull *Inter gravissimas* (Among the most important things; papal edicts, or bulls, are named after their first lines).

Most of the controversy was about calculating the *epact*, or method of intercalation for adjusting the lunar year to the solar year. Besides

(continued)

and his fellow Jesuits' brains themselves transformed by their medieval religious discipline into the observational device that verified his pendulum's timing. They synchronized their counting by the essentially medieval method of chanting—music, as part of the quadrivium, being "number in time." Similarly, Santorio's pulsilogium, converting time to length, was an Aristotelian transformation that would have been familiar to those Parisian scholars measuring their elections by the burning of a candle. Without the contributions of medieval thought, the birth of modern science would have been impossible.

this, the Gregorian calendar changes the Julian calendar in one other way: leap years are held *only* in centennial years divisible by 400. Thus, 1700, 1800, and 1900 were not leap years, and 2100 will not be—but 2000 was. This adjusts the "average" length of the calendar year to 365.2425 days, which is rather close to the current figure of 365.2421 used by astronomers.

To understand the epact and how Lilius and Clavius changed the computing of Easter, you need to remember that the full moon after the solstice is not determined observation but by a conventional calendar, with the equinox given as March 21. In the same way, the synodic months—the cycle of the phases of the moon, or *lunations*—last either 29 or 30 days (to account for the fact that the true cycle is 29.53 days). The lunations measured by the synodic calendar don't match up with the regular twelve-month tropical year measured by the regular calendar, though—it is about 11 days shorter. The Metonic cycle says the synodic and solar calendar will repeat every 19 years, assuming 19 solar years are as long as 235 synodic months. In other words, the phase of the moon will be same on a certain date as it was 19 years previously. This was the means by which Easter had been calculated up until the Gregorian reform.

Lilius and Clavius, however, saw that this was inaccurate because there is, in fact, a remainder of one day every 19 years. He therefore proposed a different system: to get the proper lunar date, one needs to add a number called the epact (from the Greek *epaktai hēmerai* or "intercalary days") to the solar date. Those to whom the lunar calendar matters keep a running count of the epact days; when the number reaches or exceeds 30, an extra or "intercalary" month is added to the lunar calendar, and 30 is subtracted from the epact. So what we have is

The Great Escapement

Although Vincenzo Galilei did not succeed in his attempt to create a pendulum clock, his father's idea came to fruition shortly thereafter. By 1656, the Dutch scientist Christiaan Huygens (1629–1695) had developed a working pendulum clock. Huygens was one of the luminaries of his age: among many other achievements, he was the discoverer of Saturn's moon Titan and an early pioneer in the field of probability.[7] By 1666, he had moved from the Netherlands to Paris to join Louis XIV's newly established French Academy. Though members of Europe's scientific community often maintained lively correspondences,

a cycle of 19 years, each having twelve or thirteen synodic months. One of these synodic months will have its 14th day—the day of the full moon—fall on or after March 21. The Sunday after this day is Easter.

The immediate effect for most people in the lands that adopted the Gregorian calendar was rather simple: the calendar skipped ahead 10 days, so Thursday, October 4, 1582, was followed by Friday, October 15, 1582. Of course, the Gregorian calendar was not adopted everywhere at once—Catholic countries such as Spain (and its vast empire) took it up immediately, but the British Empire (including America) did not adopt the Gregorian calendar until 1752. In fact, Greece adopted it only in 1923, and the Russian "October" Revolution of 1917 that brought in Communist rule actually took place in Gregorian November. The Eastern Orthodox Church still uses the Julian calendar and the Metonic cycle for religious purposes, which is why Orthodox Christmas is on January 7: the Julian calendar is currently 13 days behind the Gregorian. In 2100, the Gregorian calendar will skip a leap year, so the Julian calendar will be 14 days behind, and Orthodox Christmas will be on January 8. Orthodox Easter varies more widely, since the Eastern church still uses the Nicaean rules for calculating the date.

Gregory also made January 1 New Year's Day. Formerly, the New Year had been considered to begin at different dates in different places—you could travel from one city to another and discover you'd

(continued)

as we shall see, personal and political vendettas also played a role in scientific and technological developments. Just as France, the Netherlands, and England were rivals for power in Europe, so, too, were the members of the French and British academies often in conflict.

At first, Huygens's pendulum clock was nothing more than a variation on the old-fashioned virge-and-foliot escapement. Galileo's idea had been to allow a wheel to rotate every time a pendulum swung in one direction; the problem is that a large swing is needed to move the virge, which reduces accuracy. What Huygens did was to add gearing to make the movement of the virge much larger than the swing of the pendulum. Huygens's clock was far more accurate than any virge-and-foliot clock and could accurately run for three hours with a loss of only one second. In fact, he was able, for the first time, to track the minutes—though the minute hand

gone back or forward a year. This aspect of the calendar reform had less effect than you might think, because various places had already begun to make January 1 New Year's Day at the beginning of the sixteenth century. Again, Protestant countries were late adopters. In fact, the British Empire and its American colonies kept the "civil" New Year on March 25 until they adopted the Gregorian calendar in 1752; beforehand, it was common to write the date as, for instance, "1665/66." Some people suggest that the calendar change is how April Fool's Day started, but there's also evidence that this custom existed long before the Gregorian changeover—for instance, the pagan Roman Hilaria holiday was celebrated on March 25.

1. There are many good sources on the Gregorian calendar; see especially G. V Coyne, Michael A. Hoskin, and Olaf Pedersen, eds., *Gregorian Reform of the Calendar: Proceedings of the Vatican Conference to Commemorate Its 400th Anniversary, 1582–1982* (Vatican City: Pontifical Academy of Sciences, 1983); and Bonnie J. Blackburn and Leofranc Holford-Strevens, *The Oxford Companion to the Year: An Exploration of Calendar Customs and Time-Reckoning* (Oxford: Oxford University Press, 2003).

2. A historical note: originally, many early Christian congregations followed the Jewish observation of Passover, which is on the full moon of the month of Nisan. If you remember our first chapter, Judaism, similar to ancient Babylonian practice, uses a lunar calendar, and the Jewish calendar is so arranged that Nisan will always fall after the spring equinox. The problem was that this would often be a weekday,

was given its own clock face, running what we would now call counter-clockwise. The motive power, however, was still given by a weight being pulled down by gravity; the pendulum only regulated the mechanism of the clock. Inaccuracies were introduced by a number of factors, notably how precise the gearing could be made, friction and wear between the various moving parts, and expansion and contraction with heating and cooling. Further innovation would be required to make it reliable.[8]

Still, the pendulum was a great innovation: it was the first *harmonic oscillator* used in timekeeping. A harmonic oscillator is an object that, when disturbed from its equilibrium position, experiences a restoring force that is proportional to the displacement. For instance, when the bob of the pendulum is lifted against gravity, it gains potential energy. When it is released, the kinetic energy will carry it back through the resting point and to the other side, where the kinetic energy will again be

whereas many Christians believed that Easter ought to be celebrated on a Sunday. In 325, the Emperor Constantine, wishing to unify the various practices of the early church, commanded a church council be held at Nicaea, a city now in modern İznik in northern Turkey. The bishops and other authorities gathered at the Council of Nicaea decreed that Easter would always be the Sunday following the first full moon after the equinox. This didn't end the controversy, though—since the ultimate authority was books, not observation, Christians in different places might have held that the equinox took place on different dates. Uniformity was obtained only when Dionysius Exiguus (the same writer who came up with the BC/AD scheme) proposed a system based on the Roman (Julian) calendar and the 19-year Metonic cycle (which we discussed in chapter 1): Easter is to be celebrated on the first Sunday after the full moon after the equinox, with the equinox always considered to be March 21. (Again, it's important to remember that the "moon" used to calculate Easter was not the astronomical moon but a conventional calendar—which isn't to say Europeans weren't taking celestial observations, as Bede and other writers had exhorted them.)

converted to potential energy, and so on and so forth until the system loses energy to friction and the bob comes to rest. Like the earlier virge-and-foliot clocks, these pendulum clocks had weights to power them; the job of the pendulum was to regulate the falling of the weight.

Huygens noticed other characteristics of pendulums, as well. Lying in bed during a brief illness, he noticed that two pendulums attached to the same beam will come into what is called *antisynchronization*, with the one swinging opposite the other. Perhaps, he thought, pendulums could be used to correct one another if one went awry—such as if the clock was on a moving ship being buffeted by the waves.

Why would someone want to put a clock on a moving ship? The answer is that, although determining latitude from the stars is fairly simple, it's nearly impossible to determine longitude by any sort of natural sign. However, comparing local time to a clock that accurately keeps the time in your port of origin is an accurate, precise, and easy-to-use means of determining longitude; since the time changes one hour for every 15 degrees of longitude you move, finding your coordinates is a simple mathematical exercise. Unfortunately, Huygens's idea of using two pendulums didn't work, and a method of keeping accurate time at sea would not be invented for roughly another century. I'll discuss the development of the seagoing chronometer, which is of no small significance to the history of timekeeping, in the next chapter.

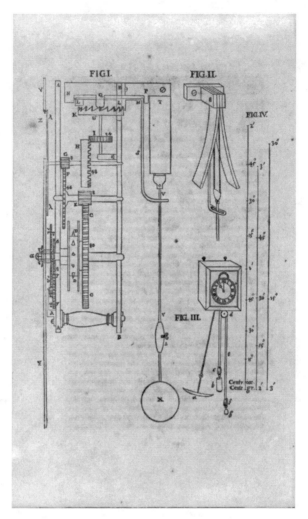

Figure 3.2. Huygens's pendulum clock.

Huygens experimented to determine the cause of the antisynchronization phenomenon. It wasn't air currents, since he found the effect still
happened if he erected a glass partition between the pendulums. Instead,
he theorized that the pendulums affected one another through the beam
that was their mutual support. The way this happens is that if pendulums are swinging out of antisynchronization, they nudge the beam
slightly to one side or the other; antisynchronization, in which these

forces cancel out, is the most energy-efficient resting state. This phenomenon of pendulums developing antisynchronization is called *entrainment*. However, entrainment happens only when the ratio of the weight of the pendulums to the clock's case falls into a certain range; if not, the forces cause one pendulum to stop while the other continues moving.

As an example of the close collaboration between scientists and craftsmen, Huygens did not build his clocks himself but contracted the work out to a clockmaker named Salamon Coster in 1657. Without skilled craftsmen to execute their ideas, the innovations and discoveries of early modern natural philosophers would have been nothing more than curiosities. There was also profit to be made by innovation: patents protected intellectual property, and inventors stood to make a tidy profit for their efforts. Making pendulum clocks would have thus been a lucrative business for both Huygens and Coster, who the scientist licensed to carry out and sell his designs.

Huygens improved his clock throughout his lifetime. In his 1673 *On the Oscillation of Clocks (Horologium Oscillatorium)*, he confirmed the observation made by earlier natural scientists such as the French priest Marin Mersenne that, contrary to what Galileo believed, a pendulum is not truly isochronous, but rather the wider the swing, the less accurate it is. Using an early method of calculus, the found that a constant acceleration under gravity requires not the arc of a pendulum's bob but rather a *cycloid*—the sort of curve a point on the rim of a wheel traces out as it rolls along. Cycloids have the unique property that a falling ball (or pendulum bob) placed anywhere on the curve will reach the end at the same time, no matter where it was placed: the greater distance the ball needs to travel compensates for the increased speed from being placed higher on the curve. To ensure accuracy, Huygens reasoned, the pendulum suspension in his clock should be of a cycloid shape, as well.

Improved Pendulum Clocks: The Anchor Escapement and Deadbeat Escapement

Another late seventeenth-century invention displaced the cycloid suspension. This was the far easier-to-build *anchor escapement*, which was invented by the great English architect and polymath Robert Hooke.[9] The anchor escapement is an improvement over the virge escapement because it uses a much smaller—and thus, more accurate—pendulum swing of 3°–4°, as opposed to up to 100° in earlier clocks. However, it has disadvantages: the pushing of the anchor on the pallets causes friction and wear. This is made worse because half the motion of the escape wheel is retrograde, called *recoil*. Though the pallets and teeth are made to minimizes this wear and tear, the accuracy of an anchor-escapement clock gets worse as it ages. The recoil also disturbs the pendulum, reducing accuracy. Plus, an anchor escapement clock cannot be moved without first stopping it, because this would cause the pendulum to oscillate and in turn cause the pallets to collide with the teeth of the escape wheel, damaging the clock. This makes it unsuitable for navigation at sea, since the rolling of a ship would ruin the clock. Finally, a pendulum with a one-second period required by such clocks needs a fairly long string, which accounts for the tall, thin design of longcase clocks (also called "grandfather clocks"), which first began appearing in the 1670s.

Some of the problems of the anchor escapement were eliminated by the *deadbeat escapement*, which was invented by the English mathematician and astronomer Richard Towneley in the 1670s, manufactured by the clockmaker Thomas Tompion, and became common by the first decades of the eighteenth century thanks to the work of Tompion's student George Graham. The deadbeat escapement improved upon the anchor escapement by eliminating the recoil.

Pendulum clocks were accurate by the standards of the day, but they weren't portable. Making a clock that could be carried around and used as a scientific instrument would require a different power source altogether. This is what led to the popularity of the spring-driven clock.

The Spring-Driven Clock, the Pocket Watch, and the Stopwatch

Although in all the clocks I've discussed before the motive force comes from a falling weight, in a spring-driven clock, the power is delivered to the mechanism by a coiled spring. (Remember that even in the case of pendulum clocks, the falling weight maintains the pendulum's movement; the pendulum, in turn, regulates the mechanism.) Because it does not require a long "drop" for a weight, a spring-powered clock can be made smaller and was the first portable mechanical timepiece, enabling the invention of the stopwatch. Time was now personal, able to be carried in one's pocket. Rather than an experimenter relying on indirect, relative measurements such as those of Nicholas of Cusa's or Galileo's clepsydrae or Santorio's pulsilogium, the spring-driven clock gave precise measurement of durations in absolute seconds.

Spring-driven clocks date back to the late Middle Ages—the first known example dates from about 1430 and was made for Philip the Good, Duke of Burgundy.[10] Centers of medieval armor production such as Nuremberg and Augsburg became centers of spring-driven clock production because the metallurgical requirements of a good spring are similar to that needed for good armor. Although spring-driven clocks achieved the goal of ease of use, they were not, at first, either precise or accurate. Medieval science couldn't really accurately control the tempering of steel, and so the quality of springs was inconsistent. Also, the force of an unwinding spring is not constant. These early spring-driven clocks were therefore admirable status symbols but not as accurate as weight-driven virge-and-foliot clocks.

To make spring-driven clocks more accurate, devices such as the *stack-feed* and *fusee* were invented to equalize the power of the unwinding spring. The desire to build a more accurate spring-driven clock led to other important innovations, as well. For instance, in the late sixteenth century, the Swiss mathematician, astronomer, and general mechanical genius Jost Bürgi, who worked in Kassel in what is now Germany, invented the *cross-beat escapement*, which improved on contemporary clocks by making two foliots travel in opposite directions to equalize the forces acting on the clock. He was also the first to use a *remontoire*, a secondary drive that moves the clock mechanism and is in turn rewound at regular intervals by the main drive. Without a remontoire, mechanical resistance to the clock's mechanism—say, snow and ice on a tower clock's hands—will affect the drive and introduce inaccuracies. In the case of Bürgi's clock, in which the main power was a spring, the hands were directly powered by the remontoire's shorter, weight-driven drive chain, which was then rewound by the spring-driven mainspring when it reached its end. This sort of indirect drive corrects for inaccuracies, because the main power source is isolated from the clock mechanism. The remontoire was critical for the continuing development of accurate timepieces.[11]

Birth of the Pocket Watch

As I mentioned above, although the desire to have a portable timepiece as a status symbol goes back to the Middle Ages, these were not particularly accurate or precise. The creation of better spring-powered mechanisms, in turn, enabled portable clocks to be made smaller, more reliable, and more accurate and precise. Specifically, the invention of the *torsion spring*, which derives its power from the uncoiling of a piece of metal, allowed clocks to be made only a few inches thick. (The torsion spring, or *mainspring*, is what gets turned when we "wind" a clock.) This innovation was (probably wrongly) credited to the German clock-maker Peter Henlein of Nuremberg (1485–1542), who was, even if he was not the inventor of the torsion spring, certainly one of the first to incorporate it into portable clocks.[12] Henlein's biography also gives us an idea of who early modern craftsmen were: he was born into the same privileged class of urban craftsmen who supported Martin Luther's Reformation and grew wealthy through their hard work and craft knowledge. He started out as an apprentice to a locksmith, thus gaining the skills he needed to join the burgeoning miniature-clock industry; later in life, he also worked on tower clocks and astronomical instruments.[13] Again, we see the confluence of patronage, craftsmanship, and the changing social and intellectual environment.

It was perhaps inevitable that these small clocks would begin being worn as fashion accessories by those who could afford them. The early ones were quite large cylinders or, later, egg shapes (the so-called Nuremberg egg) worn around the neck, affixed to the wearer's clothing, or carried in a purse. (At the time, both men and women wore purses, usually attached to their belt.) They usually told only the hour and were not very accurate, losing as much as an hour per day, but that was beside the point: they were tremendously stylish and were often made in the shape of animals, skulls (a memento mori, or "reminder of death"—a suitable idea for a clock!) or built into pomanders (decoratively pierced metal containers that held sweet-smelling spices) and other fashionable items.

The pocket watch couldn't come into being, of course, until there were pockets, which happened in the late seventeenth century. When portable clocks became smaller and more accurate, and their size, glass-covered faces, and smooth, rounded forms enabled them to be slipped smoothly into a pocket, they became no less indispensable to the gentleman of fashion than our mobile phones are to us today. The English diarist Samuel Pepys talks about acquiring a pocket watch in 1665:

> A good and brave piece it is, and [the watchmaker] tells me worth £14 [more than a laborer's annual salary]. . . . But, Lord! to see how much of my old folly and childishnesse hangs upon me still that I cannot forbear carrying my watch in my hand in the coach all this afternoon, and seeing what o'clock it is one hundred times, and am apt to think with myself, how could I be so long without one; though I remember since, I had one, and found it a trouble, and resolved to carry one no more about me while I lived.[14]

But Pepys's toy wouldn't have been really accurate and probably would have told only the hour. Early seventeenth-century watches, on average, kept pretty good time over the course of a day, but their virge mechanisms had no inherent period because, unlike a pendulum, the virge is not a harmonic oscillator. What this means is that its beat necessarily changes as the spring uncoils and the force diminishes. Despite innovations such as fusees and stackfeeds, such watches were really accurate only to within a half hour.

The great timekeeping innovation that led to accurate and portable pocket watches was the addition of the *balance spring* (or *hairspring*) to the *balance wheel*. The story of this innovation is the story of the rivalry between two men we have already met—Huygens and Hooke. Robert Hooke's genius can't really be sufficiently praised: he not only was one of the organizing forces behind the Royal Society; helped to rebuild London after the Great Fire of 1666; and published *Micrographia*, a study of the small-scale world in which he first used the term "cell" to discuss the smallest unit of a living organism; but was also a prodigious experimenter, theorist, and astronomer. He had a reputation as an irascible and jealous man; that his work is not better known is due to his professional and personal rivalries with many of his contemporaries, including Isaac Newton and Henry Oldenburg, secretary of the Royal Society. These men suppressed the knowledge of many of Hooke's innovations after the latter's death—and caused the controversy surrounding the balance spring.

The balance spring serves the same purpose in a spring-driven clock as the pendulum does in a longcase clock—it is an isochronic timing device to regulate the movement of the clock's mechanism. Balance wheels, of course, go back to the Middle Ages—the weights on the virge-and-foliot clock adjust the inertia of the crown escapement and thus the clock's timing. Galileo and Huygens thought to regulate the movement of the escapement with a pendulum, but as we saw, gravity-driven pendulums are both large and prone to disruption by movement. Thus, they are not suitable for either a pocket watch or for use on a ship. Hooke's brainstorm was to find an isochronic device that didn't rely on gravity. Putting a balance wheel together with a spiral balance spring, he found, gave quite an impressive result: every time the balance wheel is moved forward by the clock's mechanism, it is pushed back by the balance spring. The combination of the inertia of the wheel and the reliable resistance of the spring gives a constant result: the spring-and-wheel combination is a simple harmonic resonator and can accurately regulate a clock. What's more, unlike a pendulum, the device is both portable and can be made small enough to be suitable for a portable timepiece. (Note that the balance spring doesn't *power* the clock—that would be the mainspring—but works with the balance wheel to *regulate* the flow of energy into the mechanism.)

Unfortunately, Hooke didn't make any money or receive any recognition from his innovation, which he began working on in about 1658. Realizing that—thanks to the weak patent system in place at the time—anyone who improved on his ideas would get all the profits, he refused to patent the balance spring and instead tried to keep it secret, at least until he (and his collaborator, Tompion) produced a working watch. This secrecy also explains the odd way in which he published his discovery about the properties of springs, which is now called Hooke's law. To keep their findings secret, early scientists often published in code, so in 1660 Hooke released his discovery in a Latin anagram. He didn't decode the riddle for his peers until 1678, when he revealed it to be *ut tensio, sic vis* (as the tension, so the force). More simply put, Hooke's law states that the force needed to compress or extend a spring is proportional to the distance, or, mathematically, $F = kX$, where F is force, X is distance, and k is the stiffness constant of the spring.

Why did Hooke finally reveal his discovery 18 years later? Because, in the intervening period (and much to Hooke's jealous displeasure), Huygens had developed a workable balance-spring watch and claimed it as his own original invention. By 1675, the Thuret family, clockmakers to King Louis XIV, were making watches to Huygens's design in Paris and raking in a tidy profit in the bargain. Such was Huygens's faith in the mechanism that these watches lacked a fusee to equalize the power of the uncoiling mainspring. They were sufficiently accurate and precise that the Royal Society was duly impressed, and Hooke was not granted the patent he belatedly applied for. Thanks to the workable balance spring, portable timepieces could be made smaller and more accurate (though still not accurate enough for navigation), the pocket watch industry boomed, and Hooke was left gnashing his teeth.

As it turned out, Hooke's suspicions have been vindicated by history. The German-born Henry Oldenburg, the Royal Society secretary, had been carrying out the seventeenth-century equivalent of industrial espionage—transmitting Hooke's discoveries to Huygens while simultaneously hiding the records that would have given priority to the English inventor. In fact, lost minutes from the meetings of the Royal Society, only rediscovered until 2006, reveal that Hooke had a working balance-spring watch in 1670, five years before Huygens. While Oldenburg is today thought of as one of the founders of scientific peer review and an advocate for the free and open flow of information, given the weak protections of intellectual property in the late seventeenth century, not all of his contemporaries agreed with his perspective.[15]

One further improvement needed to be made before pocket watches could be really accurate: the virge escapement had to be replaced with a *lever escapement*, invented by the English clockmaker Thomas Mudge in about 1755, with several improvements made into the nineteenth century. Thanks to this invention, escapements became the most accurate they would be until the invention of electronic timekeeping in the twentieth century. The technique of their manufacture was so prized— and the English technique of using small jewels for the parts of the mechanism that encountered friction was so secret—that craftsmen were forbidden to emigrate.

Pocket watches, manufactured by hand, were both immensely fashionable and immensely expensive. Unsurprisingly, there was an active trade in their theft and pawning. The pickpocket or robber who managed to steal one from a gentleman could be beaten, sentenced to hard labor, or even hanged if caught, but the profit was worth the risk: in eighteenth-century London, a reward equivalent to as much as a year's salary for an ordinary domestic servant might be offered for the return of a stolen watch. The pickpockets could be quite brazen: some dressed in gentleman's clothes and mixed with "people of quality" at lectures, concerts, and even church to lift their watches; the man carrying the ceremonial cornet and cushion at the funeral for Charles Douglas, the Earl of Selkirk, in 1739 had his gold watch stolen right out of his pocket.[16] Later in the eighteenth century, pocket watches began to be produced less expensively, such as those with painted faces made for sailors, but at the beginning of its history the pocket watch was a rare gem indeed. All of this goes to show the importance of demand in the development of timepieces: consumers' wish to carry time in their pockets spurred the development of less expensive and more accurate clocks. Again, perceived need drove innovation.

The Stopwatch

Similar to how demand drove the development of other portable timepieces, the stopwatch came from the idea of applying the tools for ordering the universe to measuring the human world—specifically, measuring the pulse as an indicator of health. We have already seen Santorio's pulsilogium, but the invention of the balance spring made even more precise, accurate measurement possible. The first stopwatch, which was the first watch with a second hand and which could be stopped by a pressing a button, was commissioned in the 1690s by the physician Sir John Floyer (1649–1734) and made by the clockmaker Samuel Watson (fl. c. 1687–1710). Floyer took his interest in pulses to the point of obsession. In his 1707 book, *The Physician's Pulse Watch*, he wrote:

> I have for many years tried pulses by the minute in common watches and pendulum clocks, when I was among my patients. After some time I met with the common sea minute-glass [that is, an hourglass] . . . and by that I made most of my experiments. But because that was not portable, I caused a pulse watch to be made which run 60 seconds, and I placed it in a box to be more carefully carried, and by this I now feel pulses. And since the watch does run unequally, rather too fast for my minute-glass, I thereby regulate it, and add 5 or 6 to the numbers told by the watch.[17]

By 1710, though, Floyer had obtained a stopwatch that was accurate to within one second of the minute-glass. This was remarkable accuracy for the time—remember, the lever escapement wouldn't be invented until the middle of the century.

The measuring of pulses with an instrument such as a watch gave some sort of objective, scientific, and modern-seeming measurement to the practice of medicine, or, as Floyer puts it, "The frequency or quickness of the pulse . . . is capable of being numbered and consequently of being most perfectly described and communicated to others." He thus constructed elaborate tables of healthy pulses by age and sex, according to the season, and before and after meals: as with other contemporaries, he sought the mathematical basis of health. Floyer also timed respiration, which, because he suffered from asthma, was a particular interest of his. Pulse and respiration are, of course, still two of the "vital" measurements that physicians take as part of the modern ritual of medicine.

The stopwatch quickly found another use in the world of horse racing. The seventeenth and eighteenth centuries were the beginnings of scientific animal breeding, and no nation was so enthusiastic about horseracing as were the English. The ancestors of what would become the Thoroughbred horse were imported in the late seventeenth and early eighteenth centuries, and races were a matter of considerable interest to the watch-owning class. Devices to better time the races—up to a fifth of a second—appeared throughout the eighteenth century. Galileo would have loved to have possessed such a device—but, of course, they couldn't have come about if he hadn't started the ball rolling on the technological developments and perceived need that led to their invention.

Clockworks: From Automata to Factory Production

The pocket watch as an article of conspicuous consumption is nothing new: timekeeping devices have been used for display ever since the days of obelisk-sundials. In keeping with their social importance, medieval public clocks were often ostentatious, with automated figures of humans and animals, astronomical models, sound effects, and bells that played carillons. For instance, the Norwich Cathedral clock, constructed in the 1320s, had 59 automata in the forms of monks, the sun, and the moon. Another early fourteenth-century example was the clock at St. Jacques in Paris with its automated procession of the Three Wise Men. By the end of the fifteenth century, mechanical men—"Jaquemarts"—were striking the hour all over Europe. The oldest surviving component of a clock automaton is a gilded rooster from the enormous astronomical clock in Strasbourg Cathedral, which was built between 1352 and 1354. The rooster, which symbolizes Christ's Passion, can open its beak, flap its wings, and crow. The 59-foot-tall (18 meter) Strasbourg clock was replaced in the 1570s by an even more lavish version.

When spring- and pendulum-powered clocks became items of consumption, they often included automata, as well. These sorts of clocks were particularly popular in Germany, and the familiar cuckoo clock, which emerged sometime during the eighteenth century, is a modern continuation of this tradition. The device to produce the cuckoo's call has remained almost unchanged since its invention.

Clockwork could also be turned to non-timekeeping purposes, and spring drives allowed these automata to be made in miniature.[18] For instance, the entertainments at the Feast of the Pheasant, given by Phillip the Good, Duke of Burgundy, in 1454, featured automata in the form of exotic animals and figures acting out moral vignettes. Automata in churches took the form of angels ascending to or descending from heaven; frightened sinners with moving, leering devils; stood in for real animals during productions of biblical plays; and, in one case—Boxley Abbey in Kent—provided the spectacle of a mechanical Jesus that seemed to be alive. A 16-inch-high (41 cm) "robot monk" built for Philip II of Spain walks around, turns his head, raises his cross, strikes his breast, and opens and closes his mouth in a spectacle of automated prayer. The monk is today housed at the Smithsonian and is still in working order. Elly Truitt has argued that such mechanisms share with clocks the ability to, in Nicholas of Cusa's words, "enfold and unfold all things"—in other words, in their repetition of automated actions, they give the viewer the ability to see with God's eye, dramatizing historical actions in time and space that model humanity's place in the divine scheme, and then rewind history to watch it all again.[19]

Automata such as these continued to be built into the eighteenth century. The Jesuits had a particular love for clockwork nativity scenes, and there were also several mechanical animals that seemed to prove the Enlightenment idea that animals are nothing more than complicated machines. Showing the clear line between timekeeping technology and the invention of industrial machinery, Jacques de Vaucanson, who invented a "digesting duck" that seemed to eat and excrete, also invented an automatic loom. For his trouble, he was pelted with rocks by weavers who feared the loss of their jobs.

Perhaps the most amazing of all these automata are the three built by Pierre Jaquet-Droz, a Swiss clockmaker, between 1768 and 1774. Thanks to ingeniously shaped cams, the automata can, respectively, play a real organ, draw four different pictures, and write out a block of text up to 40 characters long. The machines thus recorded and played back information stored in their clockwork mechanism. Jaquet-Droz had fashioned the equivalent of read-only memory out of clockwork, creating a sort of analog computer.

These devices were more than fancy toys: they seemed to do uniquely human work in a way that was completely independent of human agency. Clockwork was more than a way of telling the hour; it was a means of automating anything of which the human mind could conceive. And, as de Vaucanson's loom shows, it was not a long leap from to using gears and cams to construct flashy but ultimately spendthrift automata to using the same tools to make useful industrial machinery that would eventually supersede human handworkers and, ultimately, human calculators. It was clockmakers' technical know-how, in no small part, that enabled the Industrial Revolution and, eventually, the computer revolution.

Significance of Technological Advancements in Timekeeping

The workable pendulum clock and the balance spring were of no small significance to the Scientific Revolution. The creation of better mechanisms meant that timekeeping accuracy had improved from a loss of some 15 minutes a day to only 15 seconds. It opened up new possibilities, such as navigators being able to determine a ship's longitude from comparing an accurate shipboard clock against local time. Only two sources of error remained that prevented this dream from becoming a reality: friction between the parts and the expansion and contraction of the mechanism with heat and cold.

On a day-to-day basis, though, clocks began to be seen as accurate representations of time. Again, we see the drive for simplicity through complexity: ironically, the more sophisticated clock drives became, the easier they were to use. To answer the question, "What time is it?" one could glance at a longcase clock or pull out a pocket watch. Such devices could, in day-to-day life, be relied on to give accurate and universally agreed-upon results. Rather than mirroring the heavens, the clock—which became seen as increasingly dependable, reliable, and abstracted from any real thing—*became* time. Likewise, living one's life by means of a clock came to be seen as normal, natural, and inevitable: workers began their labor at such-and-such an hour; Parliament convened at a certain hour. Technological progress thus led to different mental and social structures. While many of these indicators were public signals such as tower clocks, with the pocket watch, time also became *personal* in a way that the communal signals of the Middle Ages had never been. The individual was now held to make the most of his (or her) employment of time.

We can see the increased reliance on clocks in a question Charles Bellair, son of Louis XIV's court geographer, put it to Huygens in 1659:

> Allow me to ask you whether in Holland, in those places where there are several pendulum clocks, they continue very long to sound the hours together; because I have had two of them converted and put a seconds pendulum in each [literally, "of three feet and several inches" which would give a one-second swing]. I have not yet been able to keep them going together four days in a row. Not that they're very far apart, and when one checks them with sundials, one cannot see a difference even after a week; but the precision of hearing is much more sensitive than that of sight.[20]

Of course, the subject would have been of interest to the son of the royal geographer, since the difference in time between two clocks could be used to determine longitude, but Bellair's use of sundials is especially intriguing: seventeenth-century sundial makers, keeping pace with technology, sought to make instruments accurate to within the minute, but their best efforts were inferior to clocks—and, of course, clocks keep

West Meets East

We can see how unique the Western perspective of timekeeping was if we look at the Jesuits in China. As I mentioned earlier in the chapter, the Jesuits were a priestly order founded to combat the Protestant Reformation. Their special calling was education and preaching. This made them the ideal missionaries to send to China.

By far the most interesting of the Jesuit missionaries to China was Matteo Ricci. Born in central Italy in 1553, Ricci received an excellent education and studied with Clavius himself. In 1578, he sailed for the East, spending four years in India before embarking for China.

Once he arrived, Ricci devoted himself to studying the Chinese language and culture. His aim was to explain both the East to the West (he is the one who gave the great scholar Kongzi his Latin name, Confucius) and Christianity to the Chinese on their own terms and in their own language; thus, his mission met with more success than any that had come previously. Timekeeping was part of the way that Ricci did this: remember that, in the Chinese system of the world, knowledge of astronomical events was special, sacred knowledge used to regulate the court rituals on which the harmony of the world depended. Using Western knowledge to predict eclipses and construct an (admittedly imperfect) clock was one way that Ricci was able to gain access to the imperial court and convert some key members of the elite. As he wrote in one 1605 letter:

> As I, with these maps of the world, clocks, spheres, astrolabes, and other devices that I made and taught, acquired the reputation of the greatest

average hours, while solar time can vary by more than a half hour over the course of the year. Scientists found a solution to the difference between clock time and solar time in astronomical conversion tables and, later, complicated and expensive *equation clocks* that kept solar time automatically, but, for most people, the common astronomical signs that had been used since antiquity were no longer good enough. The only solution to the problem of more accurate timekeeping . . . was a better clock. Mechanics had surpassed natural signs as a means of telling time. (I will discuss the *equation of time*, on how clock time and sundial time are reconciled, in the next chapter.)

mathematician in the world, and even though I don't have any books on astrology here, with certain Portuguese nautical almanacs and collections I sometimes predict eclipses more accurately than them. And thus when I say that I don't have any books and don't want to start revising their rules, few of them believe me. I also say that, if the mathematician I was talking about were to come here, we could change our tables into Chinese letters, which I can do very easily, and undertake the task of revising the year, which would greatly enhance our reputation, widening this entry into China and making us more firmly established and freer.[1]

Ricci had a great respect for Chinese culture and innovation, but the fact that European scholars could do some things their Sinitic counterparts couldn't became a justification for assertions of cultural superiority. Jean-Baptiste du Halde (1674–1743), a French Jesuit who wrote a history of missionary activity in China (though he himself never visited), described how the novelty of European technology—most especially clocks—impressed the Chinese:

This nation, naturally proud, looked upon themselves as the most learned in the world . . . but they were undeceived by the ingenuity of the missionaries who appeared at court. The proof which they gave of their capacity served greatly to authorize their ministry and to gain esteem for the religion which they preached. . . . [T]he Jesuits, perceiving how necessary the protection of [the emperor] was to the progress of the Gospel, omitted nothing that might excite his curiosity and satisfy this natural relish for the sciences. . . .

(continued)

Newton and the Philosophy of Absolute Time

Isaac Newton's idea of "absolute" time as expressed in his 1687 magnum opus *Mathematical Principles of Natural Philosophy* (*Philosophiae Naturalis Principia Mathematica*) mirrors the idea of the clock as the independent indicator of time. Even though he built on the decades and centuries of thought that had preceded him, to understand Newton is to understand a fundamental shift in philosophy and of imagining the world.[21]

Even if Newton was a "realist" with regard to time, he was not a realist in the Platonic sense. Rather, he stood at the end of a lineage of ideas that went back to the rediscovery of atomic theory in the fifteenth century. We can thank the Italian humanist Poggio Bracciolini (1380–1459) for

> It was well known, as I have elsewhere mentioned, that what gave Père [Matteo] Ricci a favorable admission into the emperor's court was a clock and a striking watch of which he made him a present. This prince was so much charmed with it that he built a magnificent tower purposely to place it in. . . . [T]he emperor's cabinet was soon filled with various rarities, especially clocks of the most recently invented and most curious workmanship. Père Pereira, who had singular talent for music, placed a large and magnificent clock on the top of the Jesuits' church. . . . This was a diversion entirely new both for the court and city, and crowds of all sorts came constantly to hear it; the church, though large, was not sufficient for the throng that incessantly went backward and forward.
>
> All these different inventions of human wit, till then unknown to the Chinese, abated something of their natural pride and taught them not to

this: while searching for lost classical literature in a German monastery, he uncovered the poem *On the Nature of Things* (*De rerum natura*) by the Roman poet Lucretius. Poggio was more interested in Lucretius's elegant Latin than in his quite shocking content, because the poem was nothing less than a course in Epicurean philosophy. The theory it expresses—that matter is made up of tiny particles that interact with one another by natural laws—heretically removes God as the lynchpin of creation. Nonetheless, atomism found some adherents such as Bernardino Telesio (1509–1588) and the French philosopher and priest Pierre Gassendi (1592–1655), who posited that infinite space and infinite time existed before God created the universe. The historian Milič Čapek saw Gassendi's idea as a clear predecessor to Newton's in their shared belief in infinite space and time being coexistent with God— Newton, who was a devout man, believed that the existence of time and space, rather than being independent of a divine being, proved God's immanence, or permeation of the universe.[22] Gassendi's ideas were, in turn, picked up by the Royal Society member Walter Charleton (1619–1707), who published his *Physiologia Epicuro-Gassendo-Charletoniana: or a fabrick of science natural, upon the hypothesis of atoms* in 1654, which Newton read at Cambridge. Newton was also influenced by his teacher Isaac Barrow (1630–1677), who preceded him as the Lucasian

have too contemptible an opinion of foreigners; nay, it so far altered their way of thinking that they began to look upon Europeans as their masters.[2]

To du Halde, writing safely from Paris for a European audience with no firsthand of China, nothing more testified to European cultural—and religious—superiority than the ability to make clocks. His attitudes were echoed by the philosophers of the Enlightenment, and the stereotype of East Asian cultures as culturally sophisticated yet somehow childlike and retrogressive persisted as a justification for imperialism.

1. Matteo Ricci, *Lettere* (Macerata: Quodlibet, 2001), 408. Translation found at http://padrematteoricci.it/Engine/RAServePG.php/P/274010010203/L/1, accessed July 3, 2018.

2. Eva March Tappan, ed., *The World's Story: A History of the World in Story, Song, and Art*, vol. 1, *China, Japan, and the Islands of the Pacific* (Boston: Houghton Mifflin, 1914), 155–162.

Chair of Mathematics at Cambridge—a prestigious academic post later held by the late physicist Stephen Hawking. In his *Lectures on Geometry (Lectiones Geometricae)*, Barrow states that "whether things run or stand still, whether we sleep or wake, time flows in its even tenor." And, later, "For as Space is to Magnitude, so does Time seem to be to Motion; so that Time is in some sort the Space of Motion."

Similarly, to Newton, time and space are real entities that exist independently of any material objects that exist and move relative to one another—or, as he calls them, *absolute space* and *absolute time*. Time, to Newton, is the independent variable, the march of integers on the x-axis. This is familiar to us today, as it is the metric against which scientists and engineers measure the movement of spacecraft, variation of voltage, or journey of cars down a production line. In Newtonian thought, this time exists independently and passes constantly, regardless of any timekeeping device. Human instruments can only mirror this more perfect reality; only by making this basic assumption can we then analyze movement. Classical mechanics—the basis of modern physics—is based on this idea. As Newton puts it in the first book of the *Principia*:

Absolute, true, and mathematical time, of itself, and from its own nature, flows equably without relation to anything external, and by another name is called duration: relative, apparent, and common time, is some sensible and external (whether accurate or unequable) measure of duration by the means of motion, which is commonly used instead of true time; such as an hour, a day, a month, a year. . . .

Absolute time, in astronomy, is distinguished from relative, by the equation or correction of the apparent time. For the natural days are truly unequal, though they are commonly considered as equal, and used for a measure of time; astronomers correct this inequality that they may measure the celestial motions by a more accurate time. It may be, that there is no such thing as an equable motion, whereby time may be accurately measured. All motions may be accelerated and retarded, but the flowing of absolute time is not liable to any change. The duration or perseverance of the existence of things remains the same, whether the motions are swift or slow, or none at all: and therefore this duration ought to be distinguished from what are only sensible measures thereof; and from which we deduce it, by means of the astronomical equation. The necessity of this equation, for determining the times of a phenomenon, is evinced as well from the experiments of the pendulum clock, as by eclipses of the satellites of Jupiter.[23]

Notice how Newton directly ties two technological innovations—the pendulum clock and the observation of the Galilean moons of Jupiter made possible by improved telescopes—to this idea of "absolute time." While Newton's ideas are theological in origin—he sees this absolute space and time as evidence for the existence of God—we can equally see them as both coming from and supporting the possibility of accurate mechanical timekeeping. However, although he considers pendulum clocks and Jupiter's moons to be superior movements by which human beings can observe the passing of equitable time, no matter how well time may be *indicated* by these things, it is not *of* these things, because time to Newton exists independently of any moving object. Medieval Scholasticism, on the contrary, would say that the timekeeping could take place only by observing moving things—for instance, the rising and setting of the stars—because it followed Aristotle in tying the nature of time and space to the movement of objects.

Just how Newton's ideas diverged from medieval Aristotelian ideas is shown in the table.

	Aristotelian "Relative" Time	Newtonian "Absolute" Time
Ontological Reality of Time	Requires intelligent observer	Exists independently of any observer
Absolute or Relative?	Relative: Can only measure one moving thing against another	Absolute: Exists independently of any outside referent
Flow of Time	Can be sped up or slowed down, depending on what is being observed	Always flows evenly
Views of Space	Relative; requires bodies to give it meaning	Absolutely immovable and incorporeal; exists independently of bodies
Nature of Matter	Infinitely divisible (as is time)	Atomistic (indivisible beyond a certain point) and obeys natural laws (as do bodies in time)

Not everyone agreed with Newton. His friend and fellow Royal Society member John Locke, though more famed as a philosopher and political theorist, had some points of difference on whether "equable motion" could exist, whether we could know if it did, or whether time could exist without space. French scholars held with Descartes's ideas that there could be no such thing as a vacuum and that space, motion, and time were relative well into the eighteenth century. This controversy was also a matter of politics: the French government was an absolutist monarchy as opposed to Great Britain's parliamentary oligarchy, and "Newtonianism" was cognate with dangerous democratic Enlightenment impulses.[24] Nonetheless, Newton's name became synonymous with the birth of modern science. As the English poet Alexander Pope (1688–1744) writes:

Nature and Nature's laws were hid in night.
God said, Let Newton be! And there was light.

In the following centuries, Newton's idea of absolute time would come to be the standard by which the world was run.

FOUR

Navigators and Regulators

Punctuality is the stern virtue of men of business, and the graceful courtesy
of princes.
   —Edward G. Bulwer-Lytton

HISTORY LOVES WINNERS, but sometimes it's the losers who are more interesting. Take, for instance, the tragedy that befell Admiral Sir Cloudesley Shovell one autumn evening in 1707 off the rocky Isles of Scilly, which
lie 28 miles (45 kilometers) off Cornwall in the southwestern corner of
Great Britain. Shovell, commander in chief of the British fleet, was returning from attacking the French navy with a flotilla of 21 ships. Though
their mission had gone well, the British were beset with storms on the
return voyage and went badly off course. The standard route would have
taken them past the Island of Ushant (French: Île d'Ouessant), the traditional marker for the southern end of the English Channel; through the
Channel; and then up the Thames and to London. On the night of October 22 (by the Julian calendar), Shovell and his men thought themselves
safely west of Ushant. However, owing to the foul weather—and the im-
possibility of determining their exact position with navigational techniques of the day—he was actually on a collision course with Scilly.
Four ships—Shovell's flagship *Association*, the *Eagle*, the *Romney*, and
the *Firebrand*—ran aground on the rocks and quickly sank. In all, about
1,500 sailors and marines were lost, with only one crewmember from
the *Romney* and 12 from the *Firebrand* surviving. The commander was
among the dead: the bodies of Shovell and his two stepsons washed up
on a beach some 7 miles (11 kilometers) away a day later.[1]

This tragedy affected Great Britain in several ways. First, Shovell was given a state burial in Westminster Abbey and treated as a national hero. Second, as they are wont to, stories and legends grew up around the disaster. One held that Shovell washed up alive, but a beachcombing Scilly native murdered him for his emerald ring. This might have some basis in reality, since Shovell was indeed missing his ring, but he was also highly unlikely to have survived very long in the frigid water. Another, less likely legend is that a common sailor from Scilly warned Shovell that they were off course and would run aground, but the low-ranking mariner was ignored (or, worse, punished). This is plainly impossible, since all hands on the *Association* were lost and no one could have related the tale. But the fact that the story was considered credible shows that navigation at sea was reckoned more an art than a science—which brings us to the third, and more lasting result of the Scilly disaster: in 1714, Parliament offered a prize of £20,000 for anyone inventing a foolproof means of determining longitude at sea. Specifically, it offered £10,000 for a method accurate to within one degree, £15,000 for 2/3 of a degree, and the full £20,000 for a method accurate to 1/2 degree. This was an enormous sum for the time—equivalent to tens of millions of dollars in today's money, though direct comparisons are impossible.

This princely reward was still deemed a bargain by those who offered it. Seafaring was the lifeblood of nations in the early modern world, but it was fraught with danger. Ships carried gold from the New World to Spain; enslaved human beings from Africa to the New World; tea and spices from Asia to England and the Netherlands; and explorers, missionaries, merchants, colonists, soldiers, and administrators to secure their mother countries' hold on their new territories. However, for lack of a means to precisely determine a ship's position, sea voyages could be extended by weeks or months, dooming sailors to slow death by scurvy, starvation, or thirst as their captains searched fruitlessly for land. This ignorance was militarily disadvantageous, as well: needing to keep to known shipping channels, Spanish galleons could easily be intercepted by British privateers. Finally, as the case of the unfortunate Cloudesley Shovell shows, there was the ever-present danger of running aground at night or in foul weather.

All of this was for sailors' inability to determine their exact position, which requires knowing the longitude. The means by which this technical challenge came to be solved by John Harrison, a self-educated man from an obscure background, is well known: Dava Sobel explains his invention of the chronometer thoroughly and entertainingly in her book *Longitude*.[2] (The term "chronometer," meaning a really accurate clock suitable for navigation, was coined by the German academic Matthias Wasmuth in 1684.)[3] I, however, think the story is more interesting if it's told from the opposite direction—not as the heroic tale of a lone, revolutionary genius who overturned centuries of thought but as a story about hard-working experts laboring collaboratively over long years. This is, after all, the more usual means by which scientific knowledge creeps forward. In this case, the experts put their faith in a means of determining longitude that did not rely on tried-and-true astronomical observations—and, ultimately, they succeeded in their task. While the myth of the lone genius is a much more appealing narrative, it is also a misleading one. Though the chronometer represents the triumph of simplicity over complexity—and thus exemplifies our themes of precision, accuracy, and ease of use—in the end, the more informative story may not be Harrison's but that of his great opponent, Nevil Maskelyne, who championed the more complicated astronomical "lunar-distance" system.[4]

Despite the fact that the chronometer eventually replaced the lunar-distance system, Maskelyne was influential to the history of timekeeping in a way that was arguably more important: he was instrumental in establishing Greenwich mean time as the standard against which all other times were to be compared. The local time at sea or in part of a far-flung colonial empire wasn't the most important time to know; rather, what was the most important was the time in an arbitrary location back in England as indicated by the face of a clock. What's more, this time wasn't taken from looking at the sun or stars at whatever location you happened to be in, but rather it was an imaginary, "corrected" standard time—Newton's absolute time made flesh. By comparing the local time against this imaginary time, you found your position on the globe. In short, universal Newtonian time was something European colonizers projected over the whole world. The chronometer was a necessary device for keeping this time, but arguably, it was the mental concept that was more important.

This chapter will first look at the history of the longitude problem, followed by the controversy about how to solve it, before turning to how the Industrial Revolution incorporated this "practical Newtonianism" to regulate society and the far-reaching effects of this development the world over. Much like ships at sea, the world of work and production for the entire human race increasingly came to be regulated by objective, independent, mechanical indicators of time that were divorced from any human perception or natural sign. This idea of time became—albeit unevenly, with fits and starts—the time the world ran on.

The Longitude Controversy

Latitude—how far north or south you are—is fairly easy to establish, as we saw in chapter 1. You can simply take the altitude of the polestar at night or compare the altitude of the sun at noon to a chart for that date. Seafarers had been destroying their eyesight doing this since time immemorial, perhaps giving rise to the stereotype of the squinty old sailor. Longitude is much more difficult. The only variable that changes as you move east or west, as Nicole Oresme saw in his circumnavigating priests problem, is the time: go east, with the direction of the earth's spin, and you will seemingly lose time as judged against the movement of the objects in the sky; go west, and you'll gain time.

Let's look at an example. Suppose you traveled east from Madrid to Constantinople (modern Istanbul). While they are at roughly the same latitude, the difference in longitude between the two cities is about 32.7°, or about 1/11 of the earth's circumference. The sun, when sighted in Madrid, will have thus seemed to have completed 1/11 more of its daily journey through the sky than it will to an observer making a simultaneous observation in Constantinople—or if you do the math, about 2 hours and 10 minutes. (The calculations, as we'll see below, are actually rather complicated because it also depends on how far north or south you are, which is why I chose two cities at roughly the same latitude for our example.) This means that if a clock is set to local time in Madrid and brought to Istanbul—assuming it didn't stop or gain or lose time during the journey—it will be about 2 hours and 10 minutes behind local time. Assuming you have an accurate timepiece, by comparing the local time with the time kept by your clock, you can determine how far east or west you had traveled. Combined with a good map, you could both avoid navigational hazards and get some idea of how long your journey would last—no small feat in the age of sail. (It's worth noting that, today, the clocks in Istanbul are only one hour ahead of those in Madrid. Of course, neither the cities nor the earth has moved, but the *social* use of time is quite different from the astronomical *observation* of time. The reason why is a complicated intersection of politics and science. We'll look at how time zones came about later on.)

However, a clock used for navigation requires a mechanism far better than the ones available at the time of Shovell's accident: because an hour is 15 degrees, an error of only four minutes translates into one degree of latitude. This converts to a distance of 68 miles (109 kilometers) at the equator, though the distance between lines of longitude shrinks to zero at the poles as the grid system converges. A sailor reckoning his or her position by a clock that lost only six seconds a day over a 40-day journey from England to the Caribbean would be off by a full degree of latitude. A working longitude clock would have to be the very image of absolute time—Newton's idea incarnate, keeping an unchanging beat independent of any outside phenomenon. Pendulum clocks were capable of such accuracy—but, as I mentioned in the previous chapter, pendulum clocks are useless on board ship. (You can work out some longitude problems for yourself in the chapter exercise at the end of the book.)

Even before the £20,000 British prize of 1714, governments had offered rich rewards for an accurate means of calculating longitude, from Phillip III of Spain posting a reward of 6,000 ducats in 1598 to France offering 10,000 livres the year after the British announced their bounty—and scholars were trying to find solutions even before Galileo and Huygens. As with latitude, it seemed that astronomy might provide a solution—for instance, in 1499, the explorer Amerigo Vespucci deduced the longitude of Brazil by observing that Mars was 3.5° off of an expected conjunction with the moon. A German astronomer named Johannes Werner proposed in 1514 that astronomers create tables of the moon's observed distance from other celestial bodies—the so-called *lunar-distance method*. This was, however, difficult. To do so accurately required generations of detailed observations and some lengthy and complicated calculations. A really good chart of lunar angles wasn't available until the late eighteenth century, and (as we shall see) in the long run the complicated lunar-distance method simply couldn't compete with the much simpler chronometer method. Still, for centuries, people placed more faith in astronomy than in clocks, or, as Jean-Baptiste Morin, whom the French government awarded 2,000 livres in 1645 for an improved lunar-distance model, said, "I do not know if the Devil will succeed in making a longitude timekeeper but it is folly for man to try."[5]

It was Gemma Frisius, the teacher of the great mapmaker Gerardus Mercator, who first proposed using a clock to find longitude in his 1530 *On the Principles of Astronomical Cosmography (De Principiis Astronomiae Cosmographicae)*. Since the ability to make an accurate clock was well beyond sixteenth-century knowledge, Galileo proposed a solution by using the moons of Jupiter as a natural clock, but the moons were hard to see with the best telescopes of the day and next to impossible to sight from a pitching ship. Also, Jupiter is not visible for part of the year. Accordingly, Galileo's idea was shelved as impractical for navigation, though it could be used to fix the latitude of a body of land.

The efforts, both crackpot and serious, continued. A 1687 English pamphlet called *Curious Enquiries* proposed (among five other brief discourses) a method of finding longitude that seems both ludicrous and cruel to modern sensibilities. Belief in the alchemical idea of "sympathy"—that is that intertwined things can affect one another at a distance, such as a person feeling injuries done to a doll incorporating their discarded hair and fingernail clippings—was widespread in the seventeenth century. The pamphlet proposed using "powder of sympathy," which was believed to be able to heal wounds remotely when applied to the weapon that made them. In this application, a dog would be intentionally wounded and placed on a ship. At a predetermined time each day, the dog's discarded bandages would be dipped in the powder, it would yelp, the time in the home port would be known, and the longitude would be found. It is not known whether this was ever tried experimentally, but the fact that few educated people seriously believed in powder of sympathy, as well as the inconvenience of torturing dogs to find longitude, points to this being a work of satire.[6] Yet no less impractical an idea was seriously proposed by two mathematicians, William Whiston and Humphrey Ditton, who suggested anchoring a series of ships 600 miles apart all across the Atlantic to fire off signals at regular intervals. These men were not fools: Whiston had helped to pass the Longitude Act, and both were respected mathematicians, followers and successors of Newton. Yet such a mad scheme seemed more likely to them than a longitude clock, which confounded the finest minds of Europe.

Christiaan Huygens, as we already saw, applied himself mightily to the problem of a longitude clock. A timepiece known as BMP2 (that is, the second model of his *balancier marin parfait*, or "perfect marine balance") represents Huygens's ultimate attempt at keeping accurate time at sea. Dating to about 1685, BMP2 was a weight-driven clock that combined a small pendulum with a foliot.[7] The clock also incorporated some innovative antifriction rollers and a device to show the *equation of time* (which I will discuss below). However, though Huygens's perfect marine balance was an immensely sophisticated clock, it did not work sufficiently well to solve the longitude problem: the verge escapement had significant recoil, and small pendulums are not much of an improvement on large ones because the ship's motion would still unavoidably upset them.

In addition, it was found that pendulums don't run consistently everywhere in the world: in 1671, the French astronomer Jean Richer (1630–1696) performed experiments at Cayenne in French Guiana, near the equator, and found that a clock ran about 2 1/2 minutes a day behind the stars and that a seconds pendulum needed to be lengthened by 1 1/4 Paris *lines* (5/48 an inch, or about 2.6 millimeters) for it to run true.[8] This is (as Newton later explained) because the earth's gravity varies by as much as 0.5 percent between the poles and the equator, where the Earth's circumference bulges out owing to its rotation (the centrifugal force of the earth's rotation also adds some additional error). Despite the amazing accuracy of Richter's observations, Huygens, frustrated that his scheme for using pendulum clocks to find latitude had proven impossible, criticized the poor Frenchman for mistreating the devices he had sent along on the expedition.

It was clear that finding longitude by means of a chronometer would require a clock using timekeeping element different from a pendulum. However, the possible sources of error are still many: friction on bearings introduces error and causes a clock to wear out; oil used for lubrication attracts dirt and can gum up the works; an unbalanced movement creates vibration; equatorial heat and polar cold cause metal to expand and contract—something that scientists measuring brass objects with brass rulers denied but that common blacksmiths were well aware of. Added to all of these everyday concerns were the effects the movement of the ship and corrosive sea salt would have on the mechanism.

The Equation of Time and Greenwich Mean Time

The technical difficulties in creating a chronometer are why most educated people believed an astronomical method would be a better way of finding longitude. This, however, required the contributions of many researchers and scholars. The first problem was agreeing on what time to use as a standard: as I discussed in the previous chapter, solar noons are not exactly twenty-four hours apart: the *mean solar time*, as tracked by a clock, differs from the *apparent solar time*, as determined by observation (say, by using a sundial or a sextant) by as much as 15 minutes. Sidereal days are always exactly the same length—twenty-four equal hours as determined by the transit of the stars—but the moments when the sun is actually at its highest in the sky are not exactly twenty-four hours apart from one day to the next.

The reason why the apparent solar time differs from the mean solar time during the course of the year is because the earth's axial tilt and slight eccentricity of its orbit causes a minor irregularity in the sun's apparent path as seen from the earth. In fact, if you photograph the sun at noon from a particular location every day for a year, it will form a thin figure eight called an *analemma*—not to be confused with the geometrical means of constructing a sundial in the appendix. The modern word "analemma" comes from an eighteenth-century technique of correcting sundials to mean solar time. If we graph the analemma, the vertical axis is the sun's declination (that is, its maximum altitude it reaches at noon over the year), while the horizontal axis corresponds to the equation of time. (If you remember, we learned about declination in chapter 1.)[9]

The means of converting the apparent solar time to mean solar time, or vice versa, is called the *equation of time*. This is not an actual mathematical formula or algorithm; the word comes from the Latin *aequatio*, "equalizing." It is a simply a table of the difference between time as kept by the sun and time as kept by a clock, which of course mirrors the revolution of the stars. From late December to mid-April, clocks run fast

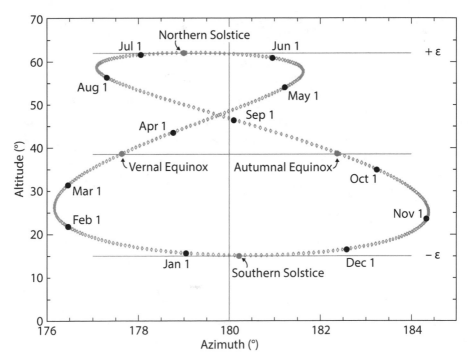

Figure 4.1. An analemma.

compared to sundials; they run slower until mid-June before seeming to speed up again; then from September to December they run behind.

Though the phenomenon of the equation of time was known since antiquity, it took on great importance in the seventeenth century as scientists and inventors sought to invent an accurate longitude clock. So-called *equation clocks*, which used gearing to show the equation of time, were produced for rich patrons in the eighteenth century (Huygens's BMP2 was an early example). However, the simplest and cheapest method is simply to follow a pre-calculated table.

The means by which the "average" day should be calculated was a matter of controversy. Huygens proposed that mean time be based on the latest sunrise. Today, though, we follow the suggestion of astronomer John Flamsteed (1646–1719), who published his tables of the equation

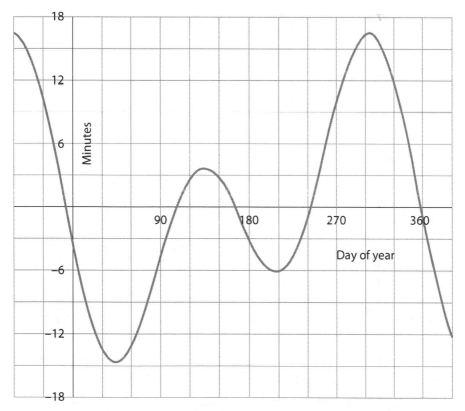

Figure 4.2. The equation of time.

of time in 1672–1673. Flamsteed proposed we use a mathematically "average" day that is between the extremes.

Flamsteed is as important a figure in the history of timekeeping and navigation as he is in astronomy, because in 1675 he persuaded King Charles II to establish the Royal Observatory in Greenwich, England, to improve the lunar-distance method—with him as Astronomer Royal and a generous salary thrown into the bargain. To this end, Flamsteed was a prolific cataloger of stars, charting more than 3,000 in his lifetime.[10] But his establishment of the observatory at Greenwich was perhaps more significant: in 1767, Maskelyne, the fifth Astronomer Royal, published his *Nautical Almanac*, containing the lunar-distance method calculating longitude from the "prime meridian" of 0° longitude, which runs through the observatory. Sailors from all nations used these indispensable tables. Maskelyne was thus instrumental in popularizing Greenwich mean time, which became the standard from which all longitude was calculated— even by the French, who held that the parallel should go through Paris.

Flamsteed was also associated with a man who would have a great effect on the development of the chronometer: Edmond Halley (1656– 1742), who first became Flamsteed's assistant and then the second Astronomer Royal after his death in 1720. Like Flamsteed, Halley came from a middle-class (which is to say a well-off but not aristocratic) background. Both were similar in that they rose through education and ability but also had the means to be introduced to the right people to further their careers. The resemblance ended there, though: Flamsteed was sober and introverted, while Halley was an extravagant, flamboyant, charismatic, brandy-drinking bon vivant.[11]

As Flamsteed's assistant, Halley helped in cataloging stars and was dispatched on expeditions to locations as far afield as the South Atlantic and Poland. Though Halley is best known today for his predictions regarding the comet that bears his name, he was also an explorer who published on weather, air pressure, and compass readings; a brilliant astronomer who helped determine the size of the solar system and was one of the first to use historical sources to predict the transit of comets; a scholar who translated ancient works on mathematics; an inventor who made a working diving bell; and a pioneer in actuarial science. His

John Harrison's Inventions

Besides the chronometer, John Harrison invented two other interesting timekeeping devices. Beginning life as a carpenter in a fairly remote part of England, he was compelled to be a jack-of-all-trades. By the age of 20, had built his first longcase clock—something that would have been impossible in London, where the trade was tightly controlled. Sometime after 1720, Harrison was commissioned to make a tower clock for Sir Charles Pelham at his estate, Brocklesby Park. This clock is still functioning after almost three centuries. Appropriately for a carpenter's son, the mechanisms of Harrison's first clocks were made entirely of wood—specifically, oak and lignum vitae, which is a hard, dense, self-lubricating wood imported from the Caribbean. This avoided the problem of needing to use lubricating oil. Besides this, we see two more innovations that would contribute to his later success in constructing a functional chronometer: the *grasshopper escapement* and the bimetallic construction of the *gridiron pendulum*.

Harrison invented the grasshopper escapement as an improvement on the anchor escapement he had made at Brocklesby Park by putting a hinge in both arms of the anchor. At first, these hinges were separate, but he later modified it so that both arms had a common pivot, making them look much like the insect that gives the device its name. As an additional advantage, the mechanism was so self-sustaining that Harrison didn't have to put free labor into maintaining it after his initial commission. One of the original wooden pallets remains in the working clock; the other was replaced only in 1880 after being damaged in an accident.

The gridiron pendulum was first used in several clocks he built with his brother James in the late 1720s. The gridiron pendulum is, simply, a means to compensate for temperature in a pendulum clock. Since metal contracts with cold and expands with heat, the length of the iron pendulum rod, and thus the period of the clock, also changes. However, different metals expand at different rates. Harrison capitalized on this observation by replacing a single rod with several rods of alternating metals such as brass and steel constructed in such a way that, as the iron rods expand and push down, the brass rods push up by the same amount. (Later iterations used zinc, which was unavailable to Harrison.) The different rates of expansion thus compensated for one another, and the clock ran true in both winter and summer. This new sort of pendulum looked like a grill or gridiron, giving the device its name. Gridiron pendulums thereafter became a standard of accuracy in nineteenth-century laboratories and factories.

measurements of compass variations were related to the longitude dilemma as well, because it was thought that one solution would be to chart the difference between magnetic north and true north (that is, between the direction a compass needle aligned with the earth's magnetic field points and the geographic North Pole). However, much of Halley's most important work was behind the scenes: it was he who financed Newton's publication of his *Principia*, and he and Newton twice gained access to Flamsteed's top-secret star charts and published them without permission—much to the first Astronomer Royal's annoyance. Though Halley's and Newton's attempts to popularize the lunar-distance method of finding longitude ultimately failed (not the least because Flamsteed burned all the copies of the star charts that he could find), Halley also championed John Harrison, who ultimately invented the chronometer. However, for a long time, it seemed as if it would be the lunar method that triumphed.

The Lunar-Distance Method versus the Chronometer

John Harrison took a long time to develop the oceangoing chronometer: between 1730 and 1757, he developed three ornate timepieces he dubbed the H1, H2, and H3. He was definitely an outsider in this race: his champion, Edmond Halley, died in 1742 and was succeeded as Astronomer Royal by James Bradley—a man wholly committed to the lunar-distance method, or as it was called, "taking lunars." A community of scientists from different nations, encouraged by the prizes offered by the British and French governments, worked together to catalog stars and describe the moon's motion against the backdrop of the night sky. Of course, to most educated people, the lunar-distance method seemed to hold greater promise than some madman's magic clock. While Harrison was tinkering in his workshop, Bradley and his assistant Maskelyne, equipped with new tables prepared by the German astronomer Tobias Mayer, had been making the observations and calculations necessary for finally taking lunars. At least in theory, sailors could use a new instrument called the octant to compare the distance from the moon to a known star, look it up in a table of ephemerides, do some equations, and thus reckon their position. (Perhaps it was inevitable that Maskelyne would be at loggerheads with the self-educated Harrison: his father had been a lawyer, and he remained an establishment man all his life, rising through the academic ranks at Cambridge and Oxford. Tying longitude to astronomy was as much a part of his personality as radical innovation was of Harrison's.)

A delay in the test voyage for the H3 due to the Seven Years' War (1756–1763) gave Harrison the opportunity to develop something completely new and more revolutionary than anything he'd produced before. Back in 1753, John Jefferies of the Company of Clockmakers had made a pocket watch to Harrison's specifications, employing not only new technological developments such as crucible steel and the use of jewels as nearly frictionless pallets but also the longitude master's own innovations such as bimetallic temperature-compensating strips and a "maintaining power" that kept the watch running during winding. Frustrated with his H3 but enamored of his incredibly accurate "Jeffries watch" and intrigued by the new technological possibilities, Harrison completed his masterpiece, the H4, in 1759 and presented it to the Board of Longitude in 1760. It is a beautiful, ornate device with metalwork recalling the astrolabes that were its medieval predecessors. It is completely different from the baroque exposed machinery of Harrison's previous attempts. Rather, it in its simplicity, it presaged the modern: the H4 is seemingly no more than a large pocket watch. Harrison was at last satisfied: "I think I may make bold to say, that there is neither any other Mechanical or Mathematical thing in the World that is more beautiful or curious in texture than this my watch or Timekeeper for the Longitude." Yet he could not claim sole credit: rather than being the singular invention of a lone genius, as the H4 relied on developments pioneered by expert craftsmen such as Jeffries.

The chronometer's sea trial was in 1761. Harrison did not go with it: He was now 68 years of age; however, his son William was in the prime of life at 33. The younger Harrison was a member of the Royal Society and London Company of Clockmakers and had spent his formative years in his father's workshop. He was the natural choice to take the H4 on a sea voyage to Jamaica.

The chronometer proved its worth early in the journey: The beer for the voyage had all spoiled, obliging the sailors to drink water, but

Figure 4.3. The clockwork of Harrison's H4. Mike Peel (www.mikepeel.net), used by Creative Commons license

William averted a possible mutiny by assuring all that they were near the Madeira archipelago and resupplies of wine. The ship's captain offered on the spot to purchase the first commercially produced chronometer. In any case, the H4, kept under lock and key save for its daily windings, performed admirably. It lost only five seconds during the 81-day outward voyage and calculated the (already known) position of Kingston to within two miles. After a week of tropical holiday, the younger Harrison and the watch were sent home. The two endured a storm-wracked journey back to England, but it was found that the H4 had lost less than two minutes in the 147-day journey.

One would think that would be the end of the matter, and the £20,000 prize, but the Board of Longitude was not yet satisfied. Bradley, the Royal Astronomer and a champion of the lunar-distance method, was not only on the board but also a contender for the prize. The conflict of interest was confirmed when William recorded in his diary how he and his father had encountered Bradley at an instrument maker's shop, and the Royal

Astronomer "in the greatest passion told Mr. Harrison that if it not been for him and his plaguey watch, Mr. Mayer and he should have shared Ten Thousand Pounds before now."[12] All sorts of reasons were invented: the chronometer wasn't accurate enough, and, worse, the younger Harrison had failed to verify the watch by the observation of the moons of Jupiter. The real reason, of course, was that in 1763, Maskelyne had published *The British Mariner's Guide*, based on the work of the German astronomer Mayer, who had finally introduced a workable lunar-distance system. Even though Bradley died in 1762 (as did Mayer), the next Astronomer Royal, Nathaniel Bliss, was no more an enthusiast of a mechanical solution to the longitude problem. Finally, it was decided that the H4 was to be put to a second trial, matched against the lunar-distance method. Harrison's masterpiece and his son were both shipped off to Barbados in the summer of 1764. Harrison objected that Maskelyne, who was sent along on the expedition, could hardly be impartial about verifying the longitude, and so the astronomer was disqualified as a judge. The H4 was the clear victor in the contest, finding the longitude of Bridgetown to within 10 miles—the four sea captains who followed Maskelyne's lunar-distance method were accurate to only within 30 miles.

If you can't win the game, change the rules: Bliss died in September 1764, and who should be appointed Astronomer Royal the following January but Maskelyne. Under his guidance, Parliament passed a new Longitude Act in 1765, giving half the prize to Harrison but withholding the other half until he turned over all his chronometers and gave detailed instructions on their manufacture. Only after duplicate chronometers had been made and tested could they could see about the rest of the longitude prize. The Board of Longitude seized all of Harrison's chronometers and papers and even forced him to give a description of how to construct it—which it then published.

Not surprisingly, Maskelyne's tests of the H4 led the Astronomer Royal to conclude that the chronometer was wanting. Meanwhile, in 1767, Maskelyne published his *Nautical Almanac*. Besides popularizing the Greenwich meridian, this continued what the Astronomer Royal had proposed to do in *The British Mariner's Guide*—to provide convenient tables for sailors to easily take lunars and find longitude at sea. It seemed that the lunar-distance advocates had won. There was only one further means of recourse: William Harrison appealed on his father's behalf to King George III. Fortunately, the king was an enthusiastic amateur scientist and had been following the longitude controversy with interest. The now 79-year-old Harrison had finished a duplicate, if plainer, chronometer, the H5, in 1770. Declaring, "By God, Harrison, I will see you righted!" in January of 1772 the king gave the watch an identical trial to the one Maskelyne had—which went badly at first because, as the king belatedly remembered, he was storing some lodestones nearby. Taken away from the source of magnetic interference, though, the H5 performed to specifications.

Thanks to the king's interest, Parliament rewarded Harrison the remainder of his money in 1773. He was further vindicated in 1775 when James Cook returned from the second of his famous voyages of exploration. As part of the chronometer's trial, Larcum Kendall, Jeffries's former apprentice, had been commissioned to duplicate the chronometer. Cook reported that—with some minor adjustments by observations of the moon—it had performed admirably.

John Harrison died in 1776; Maskelyne lived until 1811 and produced new editions of his *Almanac* until the time of his death. To be sure, the lunar-distance method had its advantages: chronometers were expensive precision instruments; octants and books were cheap. Books and celestial objects couldn't break like chronometers could. Also, the lunar-distance method seemed a natural development from finding one's position from sea and sky, while placing all of one's faith in a magic watch seemed like a poor idea. It also had its disadvantages: it required at least a half-hour's worth of calculations, was impossible when the moon wasn't visible for about six days of the month, and required the adjunct of a pocket watch for the 13 days when the sun and the moon were on opposite sides of the earth and the navigator had to rely on a star. If you needed a pocket watch anyway, chronometer proponents said, why not use a more accurate chronometer?

Even the problem of expense was not insurmountable: by the end of the 1780s, a watchmaker named John Arnold worked out how to mass-produce accurate chronometers; soon, his competitor Thomas Earnshaw worked out how to make them even more cheaply. Seafarers voted with their wallets, and though lunar tables continued to be published until the early twentieth century, the chronometer soon surpassed the lunar-distance method in popularity. After all, an accurate chronometer, though there was great genius in its making, was to the end user a simple and nigh foolproof way to determine one's longitude, and it represented the state of the art in portable timekeeping.[13] Complexity, again, was reduced to simplicity.

However, Maskelyne's influence, and that of the other establishment figures, remains. Harrison's chronometer would be useless without Greenwich mean time. This idea of standard time would be incredibly

Decimal Time

In keeping with the rationality of the new age, why not abandon the old, clunky Mesopotamian base-60 system—that is, the 60-minute hour and 60-second minute—and institute a base-10 system? After all, in everything else, we count from 0 to 10, then in tens, hundreds, and thousands. Wouldn't such a system simplify timekeeping?

There was one serious attempt to remake the hour. The French Revolution saw endeavours to reorder every other aspect of human existence, from forms of address ("citizen" instead of "monsieur") to renaming the months, so it was perhaps inevitable that the revolutionaries would try to change the time. In 1788, one would-be reformer, the lawyer Claude Boniface Collignon, suggested dividing the day into 10 hours of 100 minutes, each minute into 1,000 seconds, each second into 1,000 *tierces*, and each tierce into 1,000 *quatierces*, together with new units of measurement based on the earth's rotation. Collignon's complicated plan was never put in use, but the National Convention *did* institute a 10-hour, 100-minute "decimal" day in 1792. (It was, in case you were wondering, reckoned according to the apparent, not mean, solar day.)

However, people were too used to their traditional means of measuring time. The decimal day never caught on, and the undertaking was abandoned in 1795 with the same law that introduced the modern metric system. Nonetheless, the attempt to persuade people to use decimal time remained and had a brief revival in 1890s France. The idea has had a lasting effect, however: the one remaining field where decimal time is used today is astronomy.

influential in the coming century, guiding the world of transportation and production the world over.

Time, Industrialism, and Society

Earnshaw's mass production of the chronometer—a single example of which it took the exacting Harrison years to produce by hand—highlights just how rapidly the world was changing in the late eighteenth and early nineteenth centuries.[14] Not only were thousands of chronometers produced, but so were countless other devices, from muskets to stockings—each made by a process in which an individual performed only one task, instead of a single craftsman taking charge of significant parts of the production. The Industrial Revolution was transforming human society on a scale not seen since the invention of agriculture.

With this economic transformation came a transformation of mentality—a popular Newtonianism that affected all members of society from high to low. The Industrial Revolution depended not only on new sources of harnessing mechanical power, such as the steam engine, but also new forms of social organization, such the factory, and new philosophies of economic life, such as Adam Smith's (1723–1790) paean to capitalism, *The Wealth of Nations* (1776). The mantra of industrial capitalism was efficiency, which was applied to both mechanical production and the daily activities of life. Smith, for instance, gave the example of a pin factory: by dividing the tasks involved in manufacturing pins, each worker performed more labor, more pins were produced, and the overall operation created more wealth.

But the message was moral, as well. Christianity has an inherent temporal message: your time on earth is short, but the kingdom of heaven is eternal. In the sixteenth, seventeenth, and eighteenth centuries, new forms of faith, such as Calvinism and Methodism, equated success in this world with eternal salvation. "Waste of time is thus the first and in principle the deadliest of sins," as the great German sociologist Max Weber (1864–1920) observed in his *The Protestant Ethic and the Spirit of Capitalism*.[15] Using one's time wisely and productively, on the contrary, became a sign of industry and therefore of godliness.

The internalization of this attitude made the pocket watch the indispensable accessory of the civilized, which is to say the *time-disciplined*, man.[16] The medieval tower clock regulated society communally by means of its signals; industrial capitalism expected its subjects to regulate *themselves*. As the great economic historian David Landes (1924–2013) notes, "Punctuality comes from within, not from without. It is the mechanical clock that made possible, for better or worse, a civilization attentive to the passage of time, hence to productivity and performance."[17] The British inventor William Radcliffe, writing about the new prosperity brought about by the textile industry in his 1828 essay *Origin of the New System of Manufacture, Commonly Called Power Loom Weaving*, remarks on the improved living conditions and dress of the workers: "The men with each a watch in his pocket, and the women dressed to their own fancy," and every house was "well furnished with a clock in an elegant mahogany or fancy case."[18] Radcliffe may have been overstating the case for prosperity, but there can be no doubt that clock and watch ownership increased dramatically in the late eighteenth and early nineteenth centuries—and what should an employer traditionally give an employee upon retirement but a gold watch? Likewise, one commonplace source of the European colonizer's sense of superiority toward the colonized was the latter's supposed lackadaisical and spendthrift attitude toward the economy of time.

Western ideas of the proper way of conducting one's business were justified by an overarching secular myth: "progress." Just as medieval theologians saw human history as begun by creation and marching toward the Last Judgment and the inevitable kingdom of the saints as described in the book of Revelation, post-Enlightenment thinkers such as Herbert Spencer (1820–1903) wrote that human history was a march toward greater and greater "perfection" that would lead to happiness and prosperity for all. This could similarly be used to justify imperialism: other cultures' ways of doing things were primitive and inefficient and better replaced by Western ideas. Likewise, progress was used to justify class oppression—the differentiation of society into managers and workers was all for the best. Unsurprisingly, many early nineteenth-century industrial conflicts were rooted in capitalists' relentless imposition of mechanically ordered time upon a newly created worker class, freshly emigrated from the countryside and still accustomed to rural work patterns. Similarly, owners increasingly looked to harness "scientific" ideas to better control their workers and create more profit.

The Western cult of industrial progress as it existed in the nineteenth century had two primary temples: the factory and the railroad. The one was a center of production; the other, a means of transportation. Both affected, and were affected by, ideas of time and timekeeping. Both, likewise, working in parallel, led to important social changes. Rather than proceeding in a chronological fashion, which would be somewhat confusing, we will first look at factories and work time and how they led to the internalization of schedules and routines, and then at how advances in transportation and communication led to electric clocks and the standardization of timekeeping. Acting together, these two spheres made much of what we think of as the modern world. This mentality was, in turn, exported to the rest of the world. In the remainder of this chapter, we will consider how "standard time" was created in the early nineteenth century and exported to the colonized world in the late nineteenth and early twentieth centuries.

The Factory and Work Time

As Benjamin Franklin is held to have said, "Time is money." Nowhere is this more true than in a factory. Instead of working with the seasons or the natural rhythm of daily life, as had been traditional, economic production came to be regulated according to the relentless logic of the clock and the factory bell. Contemporaries were hardly ignorant of this tyranny: in a letter to his collaborator Friedrich Engels, Karl Marx criticized time discipline as a sort of theft. To Marx, the clock was the first applied use of the machine, "from which the whole theory of production of regular motion was developed."[19] Marx's critique of the capitalist regime was based on the idea that it "stole time" from workers—time that could be otherwise devoted to leisure activities.

As E. P. Thompson and other economic historians who have followed Marx write, labor in the premodern period was hardly performed at a uniform rate. Under the "putting-out" system that preceded the factory, entrepreneurs might deliver raw materials to the homes of their workers—say, wool to be spun into yarn.[20] Other than having to deliver their quota when their employer came around again, the workers were left to their own devices, and work could proceed at its own rhythm. Workers could engage in other profitable activities such as farm labor, spend Sundays in recreation and (often) heavy drinking, observe a "Saint Monday" of slow production at the start of the week, and speed up toward the end to earn enough for their needs and pleasures. This was an ancient practice: looking at economic records from the city of Genoa, the historian Steven Epstein has found evidence of this work pattern going all the way back to the thirteenth century.[21] Likewise, as you may remember from the second chapter, the fourteenth-century philosopher Jean Buridan noted that workers of his day gauged the passage of time by how much work they had accomplished—not that some objective measure of absolute time should guide how much work should be done.

Then came factories. The first, in the modern sense, was John Lombe's silk-throwing mill of 1721. Located on an island in the River Derwent, it was powered by water, though there was also a steam engine to keep the air warm and moist. Lombe's machines were nothing new—he copied them from Italian models—but his powering them all from one central source was. Further innovation accelerated the pace of industry: John Kay's flying shuttle of 1733 allowed cotton cloth to be woven faster, which led to James Hargreaves's spinning jenny of 1764 to spin more thread and Richard Arkwright's water frame of 1769. James Watt's steam engine—the harnessing of England's ample resources of coal to heat water and power machinery—meant that machines were no longer dependent on water sources. Thus, Edmund Cartwright's power loom, which he patented in 1785 but which did not achieve great success until the early nineteenth century, enabled cloth to be woven more quickly and cheaply than ever before.

Factories such as Arkwright's Cromford Mills were innovative in other ways, as well: workers—mostly women and children—lived on site, which gave employers a large degree of power over their employees' work and leisure schedules. Instead of completing their labor at their own pace, they were monitored and supervised at all times. This was far from new—medieval garment workers in cities and towns were often similarly constrained by clock time—but the factory led to the application of this time-work discipline to more people than ever before. Thanks to the population boom and process of *enclosure*—landlords taking over formerly public agricultural lands for personal profit—many rural dwellers moved to cities and towns to seek their livelihoods and, thus, be exposed to the idea of industrial timekeeping. The routinization of production required the routinization of people. As Landes has said, "Knowledge of the time must be combined with obedience—what social scientists like to call time discipline. The indications are in effect commands, for responsiveness to these cues is imprinted on us and we ignore them at our peril."[22]

One of the most famous examples of this system is the mills erected on the Merrimack River in Massachusetts by Francis Cabot Lowell (1775–1817). In 1814, he and several partners began the Boston Manufacturing Company. Stocked with new machines and staffed, in large part, with female workers who lived in dormitories under close supervision—and were paid less than men—the company was the first "integrated" factory in the United States. Though Lowell died three years after the factory opened, the company lived on, and by 1840, there were more than 8,000 "mill girls."[23] Life was subject to strict time discipline: the machines' production rate was measured in beats per minute; the workers labored for 12–14 hours a day, with half days on Saturday; they were subject to curfews; they were fined if they were late or did not meet quotas. Not much time was allowed for meals, education, leisure, or even personal necessity, and there were only four holidays per year. As one worker wrote in 1846:

> The operatives work thirteen hours a day in the summer time, and from daylight to dark in the winter. At half past four in the morning the factory bell rings, and at five the girls must be in the mills. A clerk, placed as a watch, observes those who are a few minutes behind the time, and effectual means are taken to stimulate to punctuality. This is

the morning commencement of the industrial discipline (should we not rather say industrial tyranny?) which is established in these Associations of this moral and Christian community. At seven the girls are allowed thirty minutes for breakfast, and at noon thirty minutes more for dinner, except during the first quarter of the year, when the time is extended to forty-five minutes. But within this time they must hurry to their boarding-houses and return to the factory, and that through the hot sun, or the rain and cold. A meal eaten under such circumstances must be quite unfavorable to digestion and health, as any medical man will inform us. At seven o'clock in the evening the factory bell sounds the close of the days work.[24]

At some mill works, such as those at Exeter, New Hampshire, employers would even wind back the clocks! Worse, an economic depression in the 1830s led to wage cuts and the employers trying to efficiently cram more and more workers onto the already overcrowded factory floors. These working conditions—and the affront to the workers' dignity and self-determination—gave rise to the first modern labor movements.

Taylorism

If we can liken the belief in efficiency to a religion, Frederick Winslow Taylor (1856–1915) certainly counts as its founding prophet.[25] From his writings came the entire idea of "scientific" management. Born in Philadelphia to a wealthy Quaker family, Taylor was expected to attend Harvard to follow in his father's footsteps as an attorney. But he defied all expectations by becoming a humble machinist at Midvale Steel Works—or not so humble, in fact, as the Clark family, which partially owned Midvale, was close to the Taylors. He rapidly rose in the ranks, becoming the foreman and chief engineer.

Taylor realized that Midvale wasn't gaining maximum efficiency from its workers. The workers were, in the jargon of the time, "soldiering"— slacking off, deliberately delaying work in the fear that fewer workers would be needed if they were all maximally productive, or simply not

Women in Horology

It's difficult not to notice that this has been by and large a story about men—male scientists, inventors, and craftsmen who created. What about women? Is the story of timekeeping a men's club?

Quite the contrary. While endemic sexism kept women out of scientific societies until relatively recently, the amateur clock historian Donn Haven Lathrop compiled a list of 750 women working in clockmaking—mostly in the eighteenth and nineteenth centuries but dating back to Agnes Dalavan, who helped to construct the tower clock in Westminster Palace in 1427.[1] Some were recorded as apprentices; some made watch parts such as cases, fusees, and chains; some kept shops running after their husbands' deaths; some were master craftswomen in their own right.

Of course, craft guilds were highly gendered organizations that didn't admit women, and commercial work was seen as a male domain. Such women were therefore somewhat rare. Yet although these 750 women made up less than 1 percent of the 300,000 names in the sources Lathrop scoured, there likely were women working in craft production who were not recorded, laboring anonymously to assist their fathers and husbands. In many ways, they were the ancestors of the unfortunate early twentieth-century "radium girls" who were poisoned painting watches with glowing radioactive material.[2] Even clockmakers in all-male shops would have been dependent on women's labor to feed them, do the laundry, and the million other small tasks of domestic labor that kept the industry running. Rather than seeing horology as an all-male profession, we need to write women back into the history of timekeeping.

1. Donn Haven Lathrop, "Women Clockmakers, Watchmakers, & Casemakers in Europe, America, & Canada 1350–1950," http://homepages.sover.net/~donnl /Women/women.html, accessed June 21, 2018. Lathrop died in 2017.

2. See Kate Moore's *The Radium Girls: The Dark Story of America's Shining Women* (New York: Thorndike Press, 2017).

using optimal processes. He made his lifelong work the study of how to encourage (or, his critics said, force) maximum effort from workers; the symbol of this philosophy—Taylorism—was the stopwatch. Rather than performing tasks in a self-directed manner, each worker was to perform them as directed, like a machine, in the most efficient way possible. As Taylor put it in his 1911 book, *The Principles of Scientific*

Management, "The best management is a true science, resting upon clearly defined laws, rules, and principles, as a foundation. And further to show that the fundamental principles of scientific management are applicable to all kinds of human activities, from our simplest individual acts to the work of our great corporations."[26]

Taylor advocated instituting this deliberate management through a system of rewards. Tasks should be optimized: through observation and experimentation, he determined that the optimal weight to be lifted with a shovel is 21 pounds. Workers should therefore be supplied with shovels that can shift this weight of whatever material must be moved. Likewise, if Bethlehem Steel wanted its workers to move pig iron most efficiently, then management should both choose those best suited for the task and arrange their schedules so that they had the optimal amount of work and rest.

Taylor's work upends the fourteenth-century logic of Jean Buridan completely: rather than workers using the amount of work they have completed to indicate elapsed time, a timekeeping machine is to dictate the amount of work to be done by the workers. This illustrates the modern move away from relative to absolute time and from human-based measurements to objective ones.

The Gilbreths

At the same time Taylor was doing his work, Frank and Lillian Moller Gilbreth were doing their own time motion studies. The Gilbreths' work, though it had different ends, goes along with Frederick Taylor's in creating the field of time and motion studies.[27]

Frank Gilbreth (1868–1924) was raised by his widowed mother. At first a poor student, he became interested in science and math but was unable to attend college and like Taylor, took up manual labor—in his case, as a bricklayer. Gilbreth quickly noticed that the bricklayers all had different methods of going about their work, and he attempted to find the best possible method. This led to the construction of a special sort of scaffold that always had the bricks within easy reach of the worker. At the same time, he was attending night school and quickly became a supervisor, then head of his own contracting company, and finally an engineer. In 1903, he met Lillian Moller (1878–1972), with whom he would have a dozen children. Moller was a very rare woman for her time in that she was both a wife and a career woman: she had a master's degree in literature and did doctoral work in psychology, eventually becoming the first PhD in industrial psychology and a professor at Purdue University.

The Gilbreths' forte was time-motion studies. For instance, they broke down the movements of the hand into discrete actions, which they called "therbligs"—"Gilbreth" spelled backward with the "th" transposed. In some ways, they were doing what fencing masters had long done—

Eadweard Muybridge and the Art of Motion

The interest in breaking down motion extended to the world of art, as well. New technology, such as photography, enabled the visualization of movement in a way that had never before been possible. The British-born photographer Eadweard Muybridge's (1830–1904) studies of "animal locomotion" began in the 1870s with a bet to see whether a racehorse has at least one hoof on the ground at all times at a gallop. (It doesn't: there's a moment of suspension.) Eventually, working at the University of Pennsylvania, Muybridge would produce more than 100,000 images of both animals in motion and humans performing various tasks such as playing sports and working. These were valued for both their scientific and their artistic value.

Among those whom Muybridge inspired was the French artist Marcel Duchamp (1887–1968), whose 1912 *Nude Descending a Staircase, No. 2*, presents its subject across time as a series of abstracted geometrical shapes. As Duchamp later recalled, "The idea of describing the movement of a nude coming downstairs while still retaining static visual means to do this, particularly interested me. The fact that I had seen chronophotographs of fencers in action and horses galloping (what we today call stroboscopic photography) gave me the idea for the *Nude*. . . . And of course the motion picture with its cinematic techniques was developing then too. The whole idea of movement, of speed, was in the air."[1]

1. Katherine Kuh, ed., *The Artist's Voice: Talks with Seventeen Modern Artists* (New York: Harper & Row, 1962), 81–93.

breaking down complex movements into simple tasks, though they were applying it to economy, not sport. Lillian applied the new science to domestic economy—modern kitchen design is very much her doing—as well as to the human element of management that they thought Taylor had neglected. The Gilbreths also applied these techniques to their own large family.

Time in the Interconnected World

Speed, movement, change: modernism not only increased the pace of manufacturing; it also connected peoples and drew the whole world into a trade network of basic commodities. Railroads, steamships, and telegraphs made the movement of humans, goods, and information across continents and oceans faster than ever before. For instance, in 1862, George Christian Giebert suggested to the chemist Justus von Liebig, who had invented a method of producing concentrated beef extract, that cattle could be more cheaply obtained in South America. By 1865, a beef extract plant had opened in Argentina, contributing to the industrialization of that country. In an era before refrigeration, the salty paste made a perfect transatlantic commodity.[28]

Living as we do today in the age of the internet, few people realize just what a world-changing invention the telegraph was. Now information could be conveyed at almost the speed of light. News of a disaster in Europe could affect markets in New York. A general could receive orders from his political leaders and send back an account of his success or failure. And, of course, astronomical observation could be transmitted as it happened. Instead of time changing gradually as one traveled from city to city, instantaneous communication made people acutely conscious of the difference in time in different places.

Railroad Time and Standard Time

The first steam-powered public railway was the Stockton and Darlington, which opened in northeastern England in 1825. Horse- and steam-powered railways had already been used for industrial purposes for some decades, but suddenly, for a moderate price, goods and people could be quickly moved across the entire country for both business and pleasure. In the United States, owing to the vast distances, railroads were built even bigger and more elaborate than they were in Europe, with the high point being the completion of the first transcontinental railroad in 1869. The results were widespread: not only was an entire tourist industry created for pleasure travel, but entire new industries sprang up as railroads brought basic commodities to massive central processing depots, such as hogs and cattle being shipped across the country to Chicago's growing stockyards.

But a railroad network requires coordination. Ever since time had started being kept by public clocks in the Middle Ages, cities and towns had used their own ideas of what time it was, which might or might not coincide with the natural time. Houston, for instance, was 33 minutes behind Chicago. Then there was railroad time: What time should a train moving between Houston and Chicago keep? Most railways kept their engineers' pocket watches synchronized to a major depot city, coordinated through the telegraph wires that were strung along train tracks. This led to quite a lot of problems: imagine if a train is scheduled to arrive at 2 PM according to railroad time—which was 1:30 local time! Worse than people missing connections or trains being late, accidents could and did occur when two trains found themselves on the same track.

The solution was to try to standardize time in far-flung places. While this would eventually have profound effects, it was first necessary to overcome deep-seated mental habits. Keeping to a factory clock was one thing; being told that sunset was at 9 PM was quite another. Time had to become divorced from its tangible indicators. And as Vanessa Ogle has written, this transformation took place at different paces and in different ways around the world.[29] People in a multicultural city such as Beirut, for instance, could be equally conscious of railway time, natural time indicators, and Muslim and Christian prayer times. The time of the colonizer was thus only one time alongside many others—though it eventually came to supplant most other timekeeping systems.

Standard Time and the First Electric Clocks

The idea of "standard time" was first brought into practice in England by the Great Western Railway in 1840. The time in London—which, in turn, was set by the Greenwich observatory and, beginning in 1852, transmitted by telegraph—became the railway's time. For a while, clocks in rail stations would display both railway time and local time, but by the middle of the 1850s, the use of standard time in public clocks had become almost universal (though it did not become law until 1880). The idea of standard time quickly spread to British-controlled India; to New Zealand, which adopted standard time in 1868; and to the United States, where railroad standard time was quickly adopted in New England after a deadly 1853 crash.[30]

Naturally, synchronizing distant clocks was a major concern of early electric clock inventors and a major motivator to electrify timepieces. The potential for such synchronized primary and secondary clocks (or, as they are sometimes called, "master" and "slave" clocks) for regulating schools, businesses, and, of course, rail systems was obvious: it would be the same time everywhere, and society would thus be ordered. However, a great deal of technical innovation would need to take place before this could be achieved—not to mention social organization.

There had been attempts to electrify clocks as far back as the turn of the nineteenth century. Some, such as the clocks made by the British telegraph pioneer Francis Ronalds, the German inventor Alois Ramis, and the Italian priest Giuseppe Zamboni in about 1815 were notably successful, if fairly unique, inventions. These early clocks used dry piles, a type of early battery that worked without the need for any liquid component. These "electrostatic clocks" would bounce a pendulum between two terminals. One such device has been ringing a bell continuously in Oxford, England, since 1840.[31]

In 1839, Carl August von Steinheil, a professor at the University of Munich, transmitted time signals from a pendulum-powered primary clock to a secondary located two kilometers away. The following year, Charles Wheatstone (1802–1875), a professor of physics at King's College, London, demonstrated a clock that used electromagnets *both* to provide impetus to move the pendulum and to synchronize other clocks—an idea he had stolen from the inventor Alexander Bain (1811–1877), who had come up with it first and demonstrated his models to Wheatstone. Bain won in court, and Wheatstone was forced to compensate him. Like Steinheil, Bain's ultimate goal was to unify time throughout the country. (Wheatstone was not entirely a fraud, though: he had earlier developed a "chronoscope," a means of using electromagnets to time physical phenomena in intervals as short as 137 microseconds.)[32]

Other inventions followed. In 1857 R. L. Jones, the stationmaster of Chester in northwest England, developed Bain's idea into a system to synchronize his secondary, key-wound clocks to the Chester main station clock by having his primary clock send out electric impulses to the bobs of his secondary clocks, which were solenoids sliding over two permanent magnets. In other words, the electric impulses did not directly drive the clocks but merely corrected the swings of their pendulums. It was not until 1872 that James Ritchie, an Edinburgh clockmaker, invented a "regulator clock" where the pendulum of the secondary clock was directly driven by the signals from the main clock. Other systems for synchronization, such as those developed by John Alexander Lund in 1876 or Ritchie in 1878, relied on the minute hand of the secondary clock being "captured" by an electromagnetic system and then released when a signal is given by the main clock, while those developed by Augustus Stroh in 1869 and Robert James Rudd in 1898 relied on affecting a small secondary pendulum.[33]

The Spread of Standard Time

In 1870, Charles F. Dowd (1825–1904), president of Temple Grove Ladies Seminary (now Skidmore College) proposed a single standard time for American railways, with different time zones drawn on natural geographical boundaries such as the Appalachian Mountains. He also proposed using Greenwich standard time as a neutral benchmark, with the first time zone calculated as 75° west of Greenwich. Not much action was taken on Dowd's idea until 1881, when William Frederick Allen (1846–1915), an engineer, railroad expert, and editor of the *Travellers'*

Daylight Savings

Perhaps it was inevitable that a New Zealander, George Hudson (1867–1946), first proposed setting clocks ahead an hour in the summer in 1895.[1] (Benjamin Franklin had jokingly proposed such a thing in a letter to the *Journal de Paris* on April 26, 1784, but nothing came of it.)[2] New Zealand, after all, experiences the date shift ahead of almost any other nation. Hudson's idea was to sacrifice normal sunrise times in favor of more light later in the day and thus both reduce energy consumption and enable outdoor recreation in the winter. The plan received much acclaim from both the British and New Zealanders, but it was never adopted. In fact, it was Britain's enemies, Germany and Austria-Hungary, that first implemented it as a wartime measure in 1916. After World War I, daylight savings time was abandoned, then readopted during World War II, and then taken up again during the energy crisis of the 1970s.

Daylight savings returns an element of the "natural" seasonal experience of time that strict clock time and rationalized scheduling removed from modern life. Unfortunately, it also goes against the trend toward simplicity: we must remember to set our clocks back or forward, and it is not consistently observed around the world. Many people also resent rising before the sun and "losing" an hour of sleep in the vernal "spring forward." Farmers and other whose work is daylight dependent also find it difficult and have been some of the largest advocates for repeal. The effects of daylight savings on health, accident rates, and the economy continue to be debated.

1. G. V. Hudson, "On Seasonal Time-Adjustment in Countries South of Lat. 30°," in *Transactions and Proceedings of the New Zealand Institute*, edited by James Hector (Wellington: John McKay Government Printing Office, 1895), 28.734.

2. "Daylight Savings Time," WebExhibits, http://www.webexhibits.org /daylightsaving/index.html, accessed August 24, 2017, is a well-researched website on this.

Official Guide to the Railways, picked up on it and proposed five standardized time zones with one-hour time differences—similar to the ones that currently exist but with the borders drawn through major depot cities such as Detroit and Buffalo. He also proposed that the standards be made international.[34]

Proposals for standardizing time met with resistance not only from towns but also from corporations and even the government. Time was a commodity in the nineteenth century: while the Signal Service, part of the Department of War, favored universal time, the Naval Observatory continued to advocate local times because of its business relationship with the Western Union telegraph service, which sent telegraph time signals to cities and ports. However, thanks to Allen's tireless campaigning, his proposal slowly gained traction and was adopted in 1883 at a meeting of the General Time Convention, a railway trade association, and put into practice on November 18, "the day of the two noons." This was purely a measure taken by private industry, though; time zones did not become federal law until 1918.[35]

Exporting Time

In the 1870s, while efforts to create standard timekeeping in industrialized nations advanced in a piecemeal fashion, advocates such as Allen and Sandford Fleming (1827–1915), the Scottish-born chief engineer of the Canadian Pacific Railway, proposed abandoning both local time and solar time in favor of an entirely artificial standard—local time zones based on a signal received by telegraph from Greenwich. Learned organizations such as the International Geographical Conference concurred that there was a need for a universal time standard, though they hesitated to force any country to adopt it.[36]

The United States, owing to its vast size, had a greater need for coordination than did any European country. In 1882, Congress authorized the president to call the International Meridian Conference. While the conference was being organized, the US railways adopted the General Time Convention, so when the representatives of 26 nations came together in Washington, DC, in 1884, the issue was really all but decided. Although there was some resistance, most representatives chose to adopt the Greenwich standard and a mean solar day of twenty-four artificial hours that begins at midnight. Though debates continued in individual European countries—the French did not adopt the Greenwich meridian until 1911—most Western nations accepted the necessity of both using universal time and coordinating social activity. Railroads, government, offices, village church bells, and peoples' daily schedules now followed not local solar time but a standardized, artificial time sundered from any natural sign and instituted from the top down.[37] The process did not proceed uniformly or smoothly—the cows in the fields might have kept their own time for milking—but it was inexorable.[38] The process of creating unified nation-states necessitated unified time.

But what about non-European nations? There is no neat narrative here: the railroad and telegraph did not instantly change life on the ground for the vast majority of people living under colonial regimes. Rather, it was only in major administrative cities that European time prevailed, and then only as one time among many. For the most part, the adoption of standard time in the colonies was something that local governors pushed for quite independently from any motive force from the home country.[39] The effect was one of influence, not of coercion: working and playing in tropical or desert climates according to a mean time separate from natural rhythms was widely seen as a notion of questionable wisdom among colonial administrators.[40] Meanwhile, people in the rest of the world followed their traditional times, sometimes coming into conflict with colonial employers' or missionaries' agendas.[41] Beirut, for instance, used both Ottoman hours and European hours, and the bell of the clock tower run by American missionaries at the Syrian Protestant College quickly came to serve the same purpose as similar devices had in medieval Europe and gave both systems a common basis.[42] The adoption of European-style universal time in the colonized world was thus, in Vanessa Ogle's words, "uneven."

It was after independence that colonized nations felt the need to imitate their former masters. Just as European customs were seen as superior, standard time had become desirable, associated with scientific modernity. In Latin America, South Asia, and elsewhere, new national governments imposed one national time together with other symbols such as flags, anthems, and religions. Even in Japan, which had never been colonized by a European power, the traditional seasonal hours, kept by rather sophisticated timepieces inspired by Western examples, were replaced with Western equal hours by the modernizing reformers of the Meiji Restoration in 1873. Even though Japan was by and large not industrialized, the European calendar and system of equal hours seemed more modern and better to many. This was not a sudden change, as Yulia Frumer makes clear, but stands at the end of a long transformation of how Japanese astronomers thought about time and timekeeping that had been sparked by contact with the West but that proceeded with its own rationality and in its own cultural context.[43]

While by the end of the twentieth century, most of the world was using a "standard offset" from Greenwich mean time (the last to adopt the standard was Nepal in 1986), the line from Maskelyne's *Almanac* to this end result was by no means a straight one. Because time zones are not set by some central authority but rather by each individual nation, the idea of time zones has also led to some interesting cultural phenomena. For instance, Detroit is halfway between the meridians of eastern and central time; it flip-flopped between the two, and also tried local mean time, until settling on eastern in 1915. As another example, the Spanish habit of dining late is well known: people in Madrid might sit down to dinner around 21:00 (9 PM), and prime-time television doesn't start until an hour later. This isn't as unreasonable as it might seem. Though they are in the same time zone, solar time in Spain is an hour later than in Naples, Italy, which is at roughly the same latitude. Spain *used* to be in the same time zone as Britain (and Portugal still is), but General Francisco Franco, the Fascist dictator of Spain, decided to switch his country to German time in the 1940s, and it was never changed back. By doing everything later, the Spanish are merely living in accordance with solar time—though they tend to lose an hour or so of sleep as a result. As sleep-deprived as the Spanish might be, western China has it worse: because the government decided in 1949 that all Chinese clocks should be set to Beijing time, they are roughly three hours behind solar time.

For our day-to-day experience, the significance of the standardization of time is this: the time we are accustomed to keeping is not natural solar time at all, which depends on when the sun reaches its azimuth at a particular latitude and longitude, but rather an artificial measure divorced from anything observed. The "average day" as observed at Greenwich, England, is the standard; the time shifts by one hour every 15 degrees of longitude. It also greatly simplifies the process of "knowing

The International Date Line

So, if there is one standard time, the question remains: At what point does the date "flip"? After all, if you circumnavigate from east to west, you gain a day; if you travel westward, you lose a day (as is observed by Phileas Fogg in *Around the World in Eighty Days*). At a certain point, a line needs to be drawn where today becomes tomorrow (or vice versa).

The international date line runs through the Pacific Ocean at roughly 180° west longitude—that is, on the opposite side of the earth from the Greenwich meridian. It isn't a straight line, of course: it swerves to avoid various islands and nations and stops short of Antarctica.[1] The line at sea is based on a 1917 agreement between Britain and France, which most major seagoing powers adopted it by 1925. However, since time zones on land are left up to individual nations, some interesting phenomena have arisen. Alaska, when purchased by the United States in 1867, both changed from one side of the date line to the other and from the Julian to the Gregorian calendar—Saturday, October 7, 1867 (Julian), to Friday, October 18, 1867. After 119 years, Samoa, which shifted to the eastern side of the date line when it became an American colony but since becoming closely tied with Australia and New Zealand changed back to the western side by "skipping" Friday, December 30, 2011. (This created problems for the island's Seventh Day Adventist population as to which day was the "real" Sabbath.) The Republic of Kiribati in effect caused the date line to "shift" eastward by having the eastern part of its territory join the western parts on the other side of the date line in 1979. Two uninhabited US territories, Howland Island and Baker Island, have the time zones of UTC −12, the latest in the world; on the other side of the line, parts of Kiribati and Samoa (in the summer) are UTC +14; New Zealand is UTC +12 in winter and +13 in summer. Kiritimati (Christmas Island), a territory of Kiribati, was where the new millennium began earliest—which made it a considerable tourist draw for New Year's Eve in the year 2000 (even if, properly speaking, the twenty-first century began in 2001).

1. NOAA's website, "What Is the International Date Line?," at https://oceanservice.noaa.gov/facts/international-date-line.html, accessed August 24, 2017, is very useful on this.

what time it is," as there is one time for everyone, regardless of location. Once standard time was adopted, the complicated calculations by observatories and resetting of timepieces by travelers became outdated as the oxcart or sail-powered schooner. This would not have been possible without modern technology such as the railway and the telegraph, but then, it wouldn't have been needed, either. Greenwich mean time—now replaced by coordinated universal time (*temps universel coordonné*, or UTC), or "Zulu" time—is a single, universally valid time reference.

FIVE

Rationalization and Relativity

"If you knew Time as well as I do," said the Hatter, "you wouldn't talk about wasting IT. It's HIM."

"I don't know what you mean," said Alice.

"Of course you don't!" the Hatter said, tossing his head contemptuously. "I dare say you never even spoke to Time!"

"Perhaps not," Alice cautiously replied: "but I know I have to beat time when I learn music."

"Ah! that accounts for it," said the Hatter. "He won't stand beating. Now, if you only kept on good terms with him, he'd do almost anything you liked with the clock . . ."

—Lewis Carroll, *Alice's Adventures in Wonderland*

Time is an illusion. Lunchtime doubly so.

- -Douglas Adams, *The Hitchhiker's Guide to the Galaxy*

SO ACCUSTOMED have we become to Albert Einstein's theory of the relativity of space and time that we often forget just how discomfited his contemporaries were by it. For roughly two centuries, the liberal regime of capitalism, nationalism, and democracy had been anchored by the Enlightenment conception of a clockwork universe governed by natural laws, understandable by savants and taken for granted by others, that were seen to govern the individual, the factory, the nation-state, and the cosmos alike. Newtonian mechanics were as predictable as positivism and progress: time is marked on the x-axis of the Cartesian coordinate system; to the unvarying beat of its constant progression, the y-axis sketched elegant curves of phenomena ranging from gravitational acceleration to population growth. These curves could, in turn, be subjected to the rational derivations and integrations of Newton's calculus, and so the mechanism of the clockwork universe was stripped bare.

However, both scientific inquiry and the disaster of the First World War proved reality too complex to be chained by such formulations. The public discomfort with an Einsteinian, relativistic world is apparent in a December 3, 1919, *New York Times* piece, published roughly a year after the end of the war, that unveiled the German Jewish physicist's new ideas to the reading public:

> Berlin, Dec. 2., Now that the Royal Society, at its meeting in London on Nov. 5, has put the stamp of its official authority on Dr. Albert Einstein's much-debated new "theory of relativity," man's conception of the universe seems likely to undergo radical changes. Indeed, there are German savants who believe that since the promulgation of Newton's theory of gravitation no discovery of such importance has been made in the world of science.
>
> When THE NEW YORK TIMES correspondent called at his home to gather from his own lips an interpretation of what to laymen must appear the book with the seven seals, Dr. Einstein himself modestly put aside the suggestion that his theory might have the same revolutionary effect on the human mind as Newton's theses. . . .
>
> "Why is your idea termed 'the theory of relativity?'" asked the correspondent.
>
> "The term relativity refers to time and space," Dr. Einstein replied. "According to Galileo and Newton, time and space were absolute entities, and the moving systems of the universe were dependent on this absolute time and space. On this conception was built the science of mechanics. The resulting formulas sufficed for all motions of a slow nature; it was found, however, that they would not conform to the rapid motions apparent in electrodynamics.
>
> "This led the Dutch professor, Lorenz, and myself to develop the theory of special relativity. Briefly, it discards absolute time and space and makes them in every instance relative to moving systems. By this theory all phenomena in electrodynamics, as well as mechanics, hitherto irreducible by the old formulas—and there are multitudes—were satisfactorily explained.

"Till now it was believed that time and space existed by themselves, even if there was nothing else—no sun, no earth, no stars—while now we know that time and space are not the vessel for the universe, but could not exist at all if there were no contents, namely, no sun, earth, and other celestial bodies." . . .

Just then an old grandfather's clock in the library chimed the mid-day hour, reminding Dr. Einstein of some appointment in another part of Berlin, and old-fashioned time and space enforced their wented absolute tyranny over him who had spoken so contemptuously of their existence, thus terminating the interview.

The *Times* correspondent's employment of the "tyranny" of the long-case clock to force the great physicist from the edge of strangeness and back into the comfortable realm of appointments and deadlines is telling: the reader must be assured that Einstein, having destroyed time in the cosmic sense, is nonetheless reassuringly governed by the same order that binds the rest of us. Einstein has threatened the neat simplicity of Newtonian physics and railroad timetables, and this must be repaired.[1]

Today, the scientific definition of time that is used the world over has been decoupled from the observation of the heavens. In fact, the invention of atomic clocks has meant that the very definition of time has become linked to an essentially arbitrary (though extremely precise and accurate) measurement. The result is a complete reversal of the age-old order of the universe: the movements of celestial objects, like the human world of work and production, are defined by timekeeping devices, instead of the other way around. The machine-driven nature of timekeeping has led to some interesting effects, for which the Y2K bug serves as an informative case study: if the system of human timekeeping breaks down, so, too, does the society that depends on such measurements.

This chapter will first look at how Einstein overturned Newtonianism before examining the timekeeping revolution of the twentieth century and how precise, accurate clocks have similarly upended the ancient logic of the heavens as the first and primary timepiece. It will then turn to at a related challenge in the development of timing devices—

measuring the speed of light. In many ways, this was the culmination of the technological development that began with the comparison of durations by water clocks and led to the development of the stopwatch.

What Does the Theory of Relativity Actually Say?

The theories of special and general relativity—and what they mean for our understanding of time—are the subject of much misunderstanding. Simply put, relativity states that space and time are not separate ideas (as Newton held) but intimately related—a continuum called "space-time." Special relativity, which Einstein published in 1905, treats the interrelation of space and time, while general relativity, which he published in 1915, deals with gravity—which it explains as the curvature of space-time—as varying with the energy and momentum of matter and radiation. The two theories are obviously related and, because of this, I will refer to them collectively as "relativity." In this section, I will attempt to explain, in plain language, what made Einstein's theories so revolutionary and how he forever changed Western ideas of time and space. However, since this is ultimately a book about timekeeping, technology, and society and not physics and astronomy (no matter how interesting the physics and astronomy are), after we've looked at these seemingly abstract and complicated ideas, I'll talk about their implications for our everyday lives.[2]

To understand Einstein's ideas, we need to begin with the idea of *inertial frames*. Your measurement of an object's velocity depends on your inertial frame—that is, how fast you yourself are going. A simple example will illustrate this: Suppose you're sitting in an airplane traveling over the earth's surface at 600 meters per second. You throw a ball up in the air and catch it a second later. To you, the ball seems to have not moved forward in any appreciable manner. However, to an observer on the ground, the ball would have traveled 600 meters. Newton held that velocities are the same no matter where you are or how fast you yourself are going; looking at it through relativistic eyes, though, the velocity of the ball depends on your inertial frame. In some ways, the idea of inertial frames, which gained currency in the late nineteenth century thanks to the work of scientists like Ernst Mach (1838–1916), was a return to the older idea that motion, and thus time, are relative. Galileo, for instance, explained the idea that all motion is relative by giving the example of dropping a weight from the mast of a moving ship; it will assume the same trajectory as if you dropped it from a tree. Newton refuted this and upheld his idea of absolute space by observing the inertia of water in a spinning bucket: even if it doesn't seem the bucket is moving from the perspective of the water, the water will still form a concave surface due to centrifugal force.

But what about when we add the speed of light to this idea of the inertial frame? This makes the thought experiment more difficult, since Einstein understood that nothing material can travel faster than light; for anything with mass to be accelerated to the speed of light would literally take infinite energy. (Also, it would violate causality, as you will see if you work out the chapter exercise at the end of the book.) No matter what the observer's inertial frame—how fast he or she is going—light will always seem to be traveling at the same rate. This phenomenon was well understood at the time, and has since been confirmed experimentally. Because of this, we refer to the speed of light as c, for "constant" or, alternately, the Latin *celeritas* (swiftness).

To preserve this special quality of light, space and time need to warp in interesting ways. As I mentioned above, we can never reach the speed of light, because it would require infinite energy—and the more energy we add and the faster we go, the more space-time warps. As a result, an observer at a significant fraction of c will see the rest of the universe "speed up": though his or her own perception of time will be normal, time outside the spaceship will seem to pass faster. Or, to put it another way, an external observer will see the ship's chronometer as running slow, while the passenger will perceive an external clock as going faster. This gives rise to the well-known "twins paradox": if you take one of a pair of 20-year-old twins and accelerate her to a significant fraction of the speed of light before returning her to Earth, she'll be greeted upon her return by her sister's grandchildren while not having aged significantly herself. Another interesting phenomenon is that the length of the near-light-speed object also gets shorter: something traveling at high proportion of c would seem to an outside observer to contract along its axis of movement. This discovery led to one of the major proofs of Einstein's theory, the redshifting of light, which we will discuss below.

One important implication of the theory of relativity is that phenomena can't really happen at the same time. We call this the *relativity of simultaneity*. In other words, we can't really speak of an event happening "at the same time" in Los Angeles and in New York, because events that seem to happen simultaneously to an observer located in Kansas won't be at the same time to an observer orbiting in a spacecraft. More concretely, if there's not enough time for light to travel from event 1 to event 2 in the time between the events occurring, then there are frames of reference in which event 1 happens before event 2 and vice versa. Another implication of this is that we can never measure the one-way speed of light, only the round-trip two-way speed of light, because the emitter has to be located at some distance from the detector.

The classic way of explaining this phenomenon first used by the gifted young American physicist Daniel Frost Comstock (1883–1970) in 1910 and repeated by Einstein himself in 1917 is a thought experiment with a train. Suppose that we have an observer standing in the middle of a train and an observer standing on a platform watching the train go by. At the moment when the observer in the center of the train is even with the observer on the platform, our witness on the train sets off a flashbulb. For the observer on the train, the light from the flashbulb will illuminate the front and back of the train at the same time. To the observer on the platform, however, the end of the train is rushing toward the photons emitted from the flashbulb, while the front is speeding away. The light will seem to hit the back of the train first. If the observer on the platform had seen the front of the train illuminated at the same time as the back, the photons would have traveled faster than the speed of light.

The time-dilation effect predicted by relativity has been a staple of science fiction. For instance, in the classic Poul Anderson science-fiction novel *Tau Zero*, a malfunctioning spacecraft is forced to travel until it actually outlasts the lifespan of the universe.[3] The tau, or τ, in Anderson's title is the time contraction factor as measured by an accelerated observer, calculated according to the formula $\tau = \sqrt{1 - v^2 / c^2}$, where v is the difference in velocity between our observer in the spaceship and the observer back home and c is, of course, the speed of light. Physicists actually prefer to use the simpler *Lorentz factor*, calculated as $1/\tau$, represented by the Greek letter gamma, γ. (The Lorentz factor is named after the Dutch physicist Heinrich Lorentz (1853–1928), who was fundamental to the mathematics of relativity.) That is, if we accelerate our observer to about 0.86 the speed of light so that her τ is 0.5, time for her passes twice as fast (1/0.5) as it does for an observer back at her home base. Obviously, as τ becomes zero as you get closer to the speed of light (as in Anderson's title), γ becomes infinite.

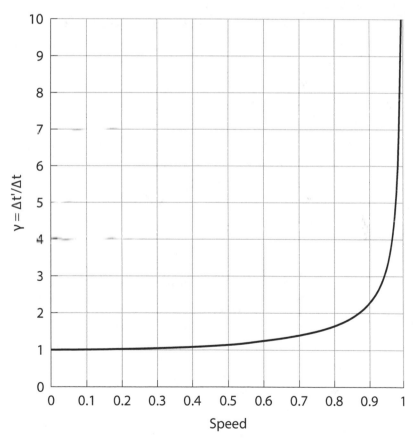

Figure 5.1. Time dilation: As speed approaches c, γ increases to infinity.

How Do We Know How Fast Light Goes, Anyway?

The idea of measuring the speed of light goes back to the dawn of modern science, and scientists' success in doing so is a terrific example of the march toward greater precision in measuring durations. Because light's swiftness is such that nothing in the universe can be used to time it in the Aristotelian sense, scientists have used some clever means to make indirect measurements.

It was obvious that light traveled faster than sound but how fast? In 1629, Isaac Beeckman attempted to find the answer by setting off a cannon and seeing the flash reflected in a distant mirror.[4] Galileo, in his *Two New Sciences* (1638), proposed a similar experiment, that two men should stand on distant hills with lanterns; when one man uncovers his lantern, the other responds to the signal by uncovering his own. Of course, no one can perform either experiment fast enough: the average human reaction speed is 0.215 seconds, so even if a cannon or a lantern-holder was standing on Mt. Everest, the light would reach the horizon (209 miles or 336 kilometers away) in about 1/1,000 of a second. Galileo's water clock would have been useless; there was no phenomenon known that could have measured the interval. Thus, both Beeckman and Galileo recorded only that there was no appreciable delay and confirmed that light travels far faster than sound.

Measuring the speed of light would require vaster distances and more indirect measurements. In 1676, the Danish astronomer Ole Rømer, working at the Paris observatory, noticed that the expected transit times of Jupiter's moon Io varied during the year. (Naturally, Rømer's work was related to the longitude project.)[5] From this and what was known of the size of the solar system, he estimated that it took 22 minutes for the light from the sun to reach the earth; shortly thereafter, Newton published the figure of "seven or eight minutes" in the *Principia*, which is closer to the correct figure of 8 minutes and 20 seconds. Newton's more accurate figure was possible because in the 1670s Giovanni Domenico Cassini (and his assistant Richer) and Flamsteed had used the parallax of Mars—that is, how much the planet's position shifts when seen against the background of distant stars from different points on the earth—to determine how far Mars is from the earth. With this figure, and knowing the relative proportions of the solar system, they figured the earth is about 87 million miles (140 million kilometers) from the sun. (The true figure is 93 million miles, or 150 million kilometers, so they were fairly close.) This gave a speed of light of 125,000 miles (about 200,000 kilometers) per second, about 75 percent of the correct figure. In 1728, the future Astronomer Royal, James Bradley, gave an even more accurate estimation, within 1 percent, by calculating the amount the earth's movement in its orbit changes the apparent angle of incoming starlight.[6]

However, all of these were indirect means of measuring the speed of light. In the nineteenth century, the French scientist Armand-Hippolyte-Louis Fizeau (1819–1896) measured the speed of light with an elegant experiment. Fizeau replicated Galileo's lantern experiment but overcame the problems of human perception by automating the timing element: a light source shone between the teeth of a rapidly rotating wheel with 720 notches in it; in place of the second lantern, Fizeau placed a mirror. It was easy to find how long light had to pass through the teeth of the wheel and return before the notch had moved; when the next tooth occluded the mirror, then the light didn't have enough time to return.[7] This, however, wasn't accurate enough, so Fizeau and his former collaborator Léon Foucault separately performed experiments that replaced the toothed wheel with a rotating mirror. This refinement determined the speed of light to within a half a percentage point (0.5 percent) of what we now know it to be. Albert Michelson (1852–1931), a Polish Jewish immigrant who became a naval officer and instructor at the Naval College in Annapolis, Maryland, improved on the Fizeau-Foucault method with more accurate equipment and achieved a result of 186,355 miles (299,909 kilometers) per second—within 0.1 percent of the current figure. For this, in 1907 he became the first American to win the Nobel Prize in Physics.[8]

As technology improved, even more precise measurements became possible. In 1950, the British scientists Louis Essen (1908–1997) and A. C. Gordon Smith used a cavity resonance resonator—a metal container that reflects radio waves—to find the speed of light as $299,792.5 \pm 1$

kilometers per second. [9] (If there was a leitmotif to Essen's career, it was precision. Remember his name, as we'll see him again in a moment when we look at the development of the atomic clock.) In 1972, a US government–backed group at the National Institute of Standards and Technology used a laser interferometer—essentially a variation of Fizeau's mirror method employing lasers—to determine the speed of light within an accuracy of 1.1 meter per second.[10] Finally, in 1983, the General Conference on Weights and Measures (Conférence générale des poids et mesures) obviated the problem by redefining the meter as "the length of the path travelled by light in vacuum during a time interval of 1/299,792,458 of a second." In other words, instead of measuring the speed of light in distance per second, the length of a meter is now set by how fast light moves in a second.

The Problems Relativity Solved

Though they are counterintuitive, Einstein's ideas about time and space resolved some important problems in late nineteenth- and early twentieth-century astronomy and physics.

One of the great tests of Einstein's ideas was explaining the precession of the perihelion of Mercury. That is, the location of the planets' perihelions—the closest points of their elliptical orbits around the sun—precess, or rotate, around the sun as time goes on. Newtonian mechanics could predict the precession of all the planets, save Mercury—the figures were estimated in the nineteenth century to be off by 38 arc-seconds per tropical century from what classical mechanics would predict. (The current estimate, using radar instead of optical telescopes, is closer to 43 arc-seconds. This isn't much; an arc-second is 1/3,600 of a degree.) Einstein's mechanics accounted for this discrepancy quite neatly because the effect of gravity is modified by the sun's mass curving space-time.

Another great test was the apparent shift of the position of stars during an eclipse. Because the sun's gravity curves light, the relative position of stars—in this case, the Hyades cluster—seen around the sun's rim during an eclipse will be different from when they are observed at night later in the year. Arthur Eddington (1882–1944) first confirmed this in 1919, and his observations did much to ratify Einstein's ideas.[11] Eddington, secretary of the Royal Society and a Quaker, was a conscientious objector to World War I, and so he was both willing to listen to a German physicist's seemingly wild idea and able to understand the mathematics involved. He was granted exemption to military service to make an expedition to the island of Príncipe, in the Gulf of Guinea, to photograph the eclipse that took place on May 29, 1919. His observations of the Hyades cluster seemed to give visual proof of relativity, and so confirmed Einstein's ideas—though there have since been accusations that Eddington was so enthused with Einstein's ideas and so eager for a postwar reconciliation with Germany through science that he misinterpreted the rather ambiguous photographic plates.[12] Nonetheless, Eddington became a great apostle and popularizer of the theory of relativity. As he versified on his 1919 experiment:

> Oh leave the Wise our measures to collate
> One thing at least is certain, LIGHT has WEIGHT,
> One thing is certain, and the rest debate—
> Light-rays, when near the Sun, DO NOT GO STRAIGHT.[13]

The final "classical" test of relativity was the redshift of light from distant stars. Simply put, because relativity changes the seeming length of observed objects, the wavelength of light, and thus the color, changes with relativistic effects. (A more technical explanation: because frequency is oscillations over time and is inversely proportional to wavelength, the wavelength, and thus the color, of light changes colors on the spectrum—goes from blue to red—as it goes from higher gravity—and more time dilation—to lower gravity and less time dilation.) The American astronomer Walter Sydney Adams first measured this phenomenon in 1925, but a truly accurate test wasn't devised until the 1950s. Today, Einstein's redshift is used to get an idea of the mass and size of distant astronomical bodies such as white dwarfs and black holes.

The Global Positioning System: A Day-to-Day Application of Relativity

The theory of relativity again underscores the relationship between time-keeping and astronomy. Besides this, an understanding of Einstein's relativity is critical to everything from planning military airstrikes to navigating supertankers on the high seas to driving a new route to school. The Global Positioning System, or GPS, is the modern equivalent of Harrison's chronometer—a technology that uses a precise knowledge of time to find a position in space. It was built by the US military at a cost of some $10 billion from 1973 to 1995 and consists of number of satellites—originally 24; today 32—placed in orbit at an altitude of 20,000 kilometers above the earth.

GPS works by these satellites continually broadcasting a time signal; when a GPS unit on the ground receives this signal, it can, by knowing the time it takes for the signal to reach the unit and the angle to selected satellites (usually three), triangulate their position. However, because of the effects of relativity, the extremely accurate atomic clocks on the satellites are 38 millionths of a second per day slower than a clock on the ground. This is because they are orbiting at about 3.9 kilometers per second, and, at their altitude, gravity is only 1/4 as strong as it is on the surface. (Objects in space still experience gravity; they orbit because they are traveling so fast they "fall" over the horizon, and the astronauts within them experience "weightlessness"—microgravity—because they are moving in the same inertial frame as their spacecraft.) This is significant enough an error that, without compensating for time dilation, the GPS system would quickly fail.[14]

Again, we see the complicated made simple: behind the easy task of using an everyday technology such as a GPS unit are some very sophisticated and difficult math and some ingenious computer programming (not to mention the difficulty of launching satellites into orbit). We'll look at the invention of the other innovation GPS satellites depend on—atomic clocks—and see how they work below.

Philosophical and Social Aspects of Relativity

As the *New York Times* article demonstrates, Einstein's relativity was prone to be spun into all sorts of philosophical ideas. After all, the world had become unraveled in the twentieth century. Modernity—belief in science, technology, the nation-state, and the march of progress toward a utopia of Enlightenment reason—had given birth to the poison gas and bombed-out cities of World War I; the Bolshevik Revolution in Russia; and, eventually, the rise of Fascism. The old order of the world was apt for questioning and overthrowing. Intellectuals could not help but see relativity as paralleling the subjectivity of twentieth-century culture, evident in everything from the rise of Freud's theories on the unconscious to the cubist and Dada movements in art to stream-of-consciousness novels such as James Joyce's *Ulysses* to jazz and ragtime and their influence on composers such as George Gershwin and Scott Joplin. Just as Newton's thought was fit into a larger paradigm, as Aristotelian thought had been before him, so, too, was Einstein placed into a worldview that was characterized by a pervasive distrust of all the things Enlightenment modernism had taken as true. While Einstein himself denied these connections (his taste in music and art were decidedly conservative), relativity seemed to say something profound to those who thought deeply about the world.[15]

Probably the most famous popularizer of relativity was the British philosopher, mathematician, writer, and general polymath Bertrand Russell (1872–1970). In his 1925 classic of science writing, *The ABCs of Relativity*, Russell first admonishes the reader that the theory does not say that perceptions of space and time are truly subjective but then goes on to write that it admits profound doubt about the universe: "Physics tells us much less about the physical world than we thought it did."[16] Russell likens us to deaf people who can comprehend sheet music in a mathematical and analytical sense but cannot hear the score. Russell cannot help himself but to extend this doubt to the march of history in general:

> The collapse of the notion of one all-embracing time, in which all events throughout the universe can be dated, must in the long run affect our

Computing the Age of the World

How old is the earth? The sun? The universe? For medieval scholars, the answer was simple: the world was created by God, and, as foretold in the book of Revelation, Christ would eventually return to judge humanity and end time. All of history was therefore bookended by creation and judgment. When exactly we could expect the Second Coming was a matter of no small import—both because it would be a poor idea to begin any long-term projects if Jesus was imminently coming back to judge humanity, and because of the potential for millennial movements (named after the thousand-year "kingdom of the saints" foretold in Revelation) springing up among the poor and oppressed who looked for an overturning of the system of the world.[1]

Early writers such as Augustine saw the history of the world as mirroring creation with a "great week" of 6,000 years followed by the 1,000-year Sabbath of the "kingdom of the saints." When, however the End of Days could come, was a moving target that could be adjusted based on the writer's particular agenda and whether he or she was using the Hebrew, Greek, or Latin text of the Old Testament. Estimations could thus vary widely: in the sixth century, Gregory of Tours put creation in 5500 BCE; a century later, Bede put it in 3952 BCE. The Hebrew calendar was closer to Bede, placing creation in 3761 BCE. Such calculations persisted into the Scientific Revolution: the Anglo-Irish bishop James

views as to cause and effect, evolution, and many other matters. For instance, the question whether, on the whole, there is progress in the universe, may depend upon our choice of a measure of time. If we choose one out of a number of equally good clocks, we may find the universe is progressing as fast as the most optimistic American thinks it is; if we choose another equally good clock, we may find the universe is going from bad to worse as fast as the most melancholy Slav could imagine. Thus optimism and pessimism are neither true nor false, but depend upon the choice of clocks.[17]

Those who looked to an imagined past for justification condemned Einstein, who left Germany in 1932, the year before Hitler became führer and Jews were stripped of academic posts in German universities with the Nuremberg laws. The Nazis burned Einstein's books and put

Ussher (1581–1656) famously added up the "begats" in the King James Bible and decided the world began on Saturday, October 22, 4004 BCE, and so could be expected to end in the year 2004. (Obviously, it didn't.) Kepler and Newton also engaged in eschatological calculations and both came up with around 4000 BCE. No matter what the date, the general consensus was that the earth was, at most a few thousand years old.

All of this changed in the eighteenth century, when people began to take note of evidence that the world was much older than had been previously thought. The study of geology utterly changed how scholars thought about time; it was plain that humanity stood at the end of an incalculably long natural history. Exactly how long, however, was a matter of debate: the British physicist and inventor William Thomson (1824–1907), who would become Lord Kelvin and after whom the kelvin scale is named, estimated that, based on how long it would have taken the earth to cool from its original molten composition to its present state, the planet was between 20 million and 100 million years old, probably younger than older. In fact, one of the greatest arguments against Darwin's explanation of evolution with the theory of natural selection was that the earth couldn't possibly be old enough to explain all the species found on it. In 1897, Kelvin revised his earlier figures downward and said the earth "was more than 20 and less than 40 million year old, and probably much

(continued)

a price on his head, and Nazi propaganda portrayed him as a traitor to Germany (he had renounced his German citizenship in 1896 and became a Swiss national in 1901), a greedy materialist, and (ironically) a Communist agitator. (Einstein was, in fact, a pacifist and active in left-wing causes.) Relativity, according to the Nazis, was "Jewish" physics and led inevitably to the decline of government, the state, and good morals. When the Nazis published *One Hundred Authors against Einstein* in 1931, the physicist wryly remarked that defeating his theories didn't require 100 writers—just a single one who was right.

Einstein reintroduced the idea of relativity—not precisely the same relativism as Aristotle but nonetheless a worldview that holds that time and space exist only in relation to moving objects. As he was quoted in the *Times* interview, "Till now it was believed that time and space existed by themselves, even if there was nothing else—no sun, no earth,

nearer 20 than 40." What Lord Kelvin didn't know about were radioactive decay and the convection of the earth's mantle. Today, we know from the rates of decay of radioactive materials found on earth, in moon rocks, and in meteorites that the earth, and the solar system, formed around 4.54 billion years ago.[2]

But what of the universe? This is a little more difficult to determine, and it depends on what mathematical model of cosmological expansion we use. For a long time, it was assumed that the universe was eternal and relatively unchanging. Kelvin's discovery of entropy in the nineteenth century meant that if the universe were infinitely old, everything would be the same temperature—which is manifestly untrue. Still, lacking better explanations, the idea of a "steady-state" universe persisted. Einstein thought that the universe was unchanging and added a cosmological constant to his equations to account for this; Eddington and others, however, thought the universe must be either expanding or contracting. Physical observation backed up the math: in the 1930s, Edwin Hubble (1889–1953) observed distant nebulae and realized that they were, in fact, galaxies—galaxies that were both very far away and moving away from us very quickly. From this evidence, it became possible to theorize the Big Bang and determine the age of the universe—about 13.8 billion years.[3] The discovery of microwave background radiation left over from the Big Bang confirms this estimate.

1. On computing the end times, the standard work is still Norman Cohn, *The Pursuit of the Millennium: Revolutionary Millenarians and Mystical Anarchists of the Middle Ages* (New York: Oxford University Press, 1990).

2. See Joe Burchfield, *Lord Kelvin and the Age of the Earth* (Chicago: University of Chicago Press, 1975). On Kelvin's age of the earth, see pp. 94 and 140.

3. Hubble, however, did not believe the universe was expanding; his opinions changed through his life. See Gale Christianson, *Edwin Hubble: Mariner of the Nebulae* (New York: Farrar, Straus and Giroux, 1995).

no stars—while now we know that time and space are not the vessel for the universe, but could not exist at all if there were no contents, namely, no sun, earth, and other celestial bodies." This, in turn, suggested that a more precise and accurate measurement of time would need to be decoupled from celestial observations.

Toward Greater Precision

The GPS system is a good example of our overall theme of ever-greater sophistication yielding ever-greater precision, accuracy, and ease of use. Timepieces with an accuracy unimaginable a century ago are now commonplace. This accuracy has permitted scientists to redefine what time is: instead of time being measured by the relative movement of physical things, the physical dimensions of objects are measured by time. In this next section, I will discuss the technical innovations of the twentieth century.

Electric Power and Timekeeping

By the early twentieth century, electric inventions were reshaping society. Streetlights, rail cars, and countless home devices were powered by this new technology. Clocks were no exception. We have already seen the application of electricity to timekeeping in regulator clocks for railway and industrial use. The expansion of this idea ultimately led to the invention of the atomic clock and the resulting redefinition of the essence of time.

Electrical clocks fall into several main categories, though the exact number of mechanisms that have been invented is so vast that describing them would take a long—and, no doubt, extraordinarily dull—book in itself. Rather than detail every electrical mechanism that has been applied to timekeeping, I will instead give an overview of the most important types and leave detailed descriptions for the glossary. The most traditional of electric clocks are mechanical clocks that simply use electricity to wind the spring, raise the weight, push the pendulum, or power a remontoire. *Synchronous clocks* rely on the power grid itself, using the frequency of the alternating current to keep time. In essence, the synchronous clock is a sort of "secondary" clock, where the "primary" is the power company's current.[18] Because tuning forks vibrate at known frequencies, a *tuning-fork* clock can be used for precise time measurements.[19] The first was created in 1866 by Alfred Niaudet-Bréguet, who used the frequency of an electrically driven tuning fork to control a small escape wheel. Fizeau used a precise clock built on this principle in his experiments measuring the speed of light. Almost a century later, Max Hetzel of the Bulova Watch Company employed same idea in the company's groundbreaking Accutron watches. Finally, *electronic clocks* do away with the mechanical system altogether. Instead, they use an electric circuit itself as the timekeeping element.

Electronic Clocks: Electronic Oscillators and Quartz Clocks

In the late 1890s, Guglielmo Marconi (1874–1937) found a way to transmit radio waves, recently discovered by the tragically short-lived German inventor Heinrich Hertz (1857–1894), over long distances. Marconi's radio waves were produced by *electronic oscillators*, electronic circuits that—much like other oscillators we have seen, such as the balance wheel and spring system and pendulums—alternate between two states in a given time. The difference between a mechanical oscillator such as a pendulum and an electromagnetic oscillator is that the latter can produce an alternating signal anywhere on the electromagnetic spectrum—in the audible range or as high-frequency radio waves. The frequency of an oscillator (or of electromagnetic radiation) is measured in hertz (Hz), a unit named after its discoverer. One hertz is one cycle per second; the shorter the wavelength, the more energy. If you count the cycles per second, then you're using the circuit as a timekeeper.

While today oscillators of various sorts are used in everything from synthesizers to amplifiers to computers, scientists in the early twentieth century were most interested in their use in radio transmission. Scientists searching for a better oscillator to use in radios first came up with the vacuum tube and later hit upon using crystals, especially quartz, as oscillators. Quartz is silicon dioxide, a common and naturally occurring mineral. Because of its molecular structure, it exhibits something called *piezoelectricity*: it will accumulate an electric charge when put under mechanical stress. Similarly, it will deform when placed in an electric field. The piezoelectric effect was discovered by Pierre Curie (1859–1906), husband of Marie Curie. The American physicist and engineer Walter Guyton Cady (1874–1974) developed the first quartz oscillator to be used in radios in 1921; George Washington Pierce (1872–

1956), a professor at Harvard, patented a circuit using a quartz crystal two years later.[20]

It was quickly realized that the frequencies of quartz oscillations could be more accurate for timekeeping than even pendulum clocks or tuning forks. In fact, the piezoelectric effect made it possible to entirely do away with mechanical elements in a clock; now, a circuit itself could be the timepiece. Furthermore, because quartz does not expand or contract much with heat and cold, it is relatively temperature insensitive. In 1927, Warren Marrison and J. W. Horton built the first quartz clock.[21] However, the reliance on bulky vacuum tubes for the electronic components limited their usefulness outside laboratories until the 1960s, when transistors and computer chips made them practical for everyday use.

The Electronic Wristwatch

Wristwatches were initially a feminine dress accessory—men, of course, had pockets in their clothes and so carried pocket watches. However, because of their utility in synchronizing troop movements during such conflicts as the Boer War, wristwatches crossed the gender divide by the end of the nineteenth century and became popular among army officers. After World War I, when coordinating assaults during trench warfare was of such critical importance that they were given to troops as standard-issue equipment, wristwatches became common for male civilian dress. It was therefore seen to be necessary that they be made accurate in a way that was not considered important when they were worn by women. However, the engineering difficulties in making a truly accurate wristwatch are many: because it moves in all different directions with the wearer's hand, this tends to upset an all-mechanical movement and limits its accuracy—thus the "synchronizing watches" scenes in old movies. Very accurate wristwatches were expensive luxury items until after World War II.

The first electric wristwatch was the Hamilton Electric 500, introduced in 1957. It was powered by a battery, which obviated the need for winding, but still used a mechanical balance wheel. The first all-electronic wristwatch was the Accutron, invented by Max Hetzel of the Bulova Watch Company. It debuted in 1960, used a 360-hertz tuning fork, and was powered by a tiny transistor. Rather than ticking, it hummed, and it was far more accurate than any mechanical watch—to a minute per month, or two seconds per day. It also had the advantages of being cheaper than watches that incorporated mechanical movements and of not being upset by movement.[22] The watches and clocks on the Apollo missions were Accutrons, since NASA did not know how a mechanical movement would work in low gravity—though the actual astronauts who landed on the moon wore Omega Speedmasters, since the Bulovas were not certified as dust-proof.

However, the Accutrons were quickly superseded in both accuracy and price point by quartz crystal watches. The first prototypes of quartz watches—the Swiss consortium Centre Electronique Horloger's Beta 1 and the Japanese firm Seiko's Astron—were introduced in 1967; the Japanese beat the Swiss to market two years later. Because the frequency at which the quartz in a clock resonates depends on how it is cut and how the electrodes are placed, the crystal must be cut precisely. Usually, in modern quartz watches, the crystal is designed in the shape of a tuning fork meant to oscillate at 32,768 hertz. An inexpensive digital counter converts this pulse to seconds. Timekeeping was entirely freed from the inaccuracies introduced by mechanical devices.

The Astron was accurate to 15 seconds per month and led to a revolution in timekeeping: watches were now less expensive and more accurate than ever before. (Of course, they could still be luxury items, as well: The Hamilton Pulsar, introduced to the public in 1972, was the first watch to do away with mechanical hands and have a digital display and cost $2,100, or more than $13,000 in 2020 dollars.) By the 1980s, quartz dominated the market for all sorts of timepieces, and the great Swiss watchmaking firms, which chose to focus on mechanical watches, lost ground to less expensive watches made in Asia. For a nominal price, it is possible today to own a timepiece that exceeds even Harrison's H4 in accuracy. Nowadays, it is even difficult to find replacement parts for a mechanical watch, and most sold today are made as cheap disposable items. The well-known Swatch, for instance, was introduced by a Swiss consortium in the 1980s as a colorful fashion accessory, but its case is plastic and not intended to be opened for repair. The wristwatch fell somewhat out of favor when cell phones became popular but are making something of a comeback as wrist computers such as the Fitbit and Apple Watch—though timekeeping is only one of these devices' functions.

Redefining the Second

The amazing precision achievable by twentieth-century science and technology, not to mention Einstein's theories, led to a need to redefine what, exactly, time was. Again, this movement came from astronomy: contrary to what had been believed for thousands of years, the rotation of the earth varies in the short term and is slowing down in the long term. How accurate can a clock be, then, when the traditional measure of time is based on astronomical observations, and when increasing precision of observation has determined that these seemingly eternal standards themselves vary? As the Dutch astronomer Willem de Sitter (1872–1934) stated the problem in 1927, "The 'astronomical time,' given by the earth's rotation, and used in all practical astronomical computations, differs from the 'uniform' or 'Newtonian' time, which is defined as the independent variable of the equations of celestial mechanics."[23] The micro-measurements of modern science require precision above all else, and it was plain that a new and more accurate definition for time would have to be found.

This need is why, in 1952, the International Astronomical Union approved the use of the *ephemeris year* at its meeting in Rome. Hours, minutes, and seconds are now defined as fractions of the earth's journey around the sun, reckoned from Greenwich mean noon on December 31, 1899 ("0 January 1900") and corrected according to a formula developed by the American astronomer Gerald Maurice Clemence (1908–1974). This date is considered to have begun an *epoch*—the reference point from which time is counted. (The current epoch used in astronomy began at approximately midnight on January 1, 2000.) In 1960, ephemeris time was adopted by the General Conference on

Weights and Measures, which is in charge of the metric system—or, as it is known today, the International System (Système internationale d'units). But this standard would itself prove ephemeral: A new invention would soon give an entirely novel standard of timekeeping—one more precise than any that had come before.

Birth of the Atomic Clock

As a graduate student at the National Physical Laboratory in Teddington, UK, Louis Essen had worked on the properties of electrical circuits. Under the tutelage of his supervisor, David Dye, he soon became well versed in creating precise quartz clocks. By cutting a quartz crystal into a ring and suspending it by six threads, he was able to make a precise oscillator that could control frequency to exacting standards. These Dye-Essen rings quickly became the basis for high-quality scientific clocks.

But there was still room for improvement. On a 1950 trip to the United States, Essen learned about an idea first suggested by the Austrian Jewish scientist Isidor Rabi (1898–1988), who had been awarded a Nobel Prize in 1944 for his discovery of magnetic resonance. In 1922, the German scientists Otto Stern and Walther Gerlach had discovered the spin of electrons with an elegant experiment that stands as a classic in the field of quantum mechanics. Electrons, as you may remember from basic chemistry, are negatively charged particles that exist in a cloud around the nucleus of atoms; each has the properties of spin and orbital motion. Because of this movement, each atom has momentary electric charge. Using this phenomenon, Stern and Gerlach were able to use a magnetic field to split a beam of vaporized silver atoms into two separate streams. In series of experiments he conducted at Columbia University in the 1930s, Rabi and his team had discovered that he could use radio waves to "tune" atoms and change their charge. Furthermore, he noted that different atoms change state—absorb and reemit energy—at different, predictable frequencies, depending on the element and the frequency of the radio wave. In short, atoms could be made to act as perfect natural oscillators.

Rabi suggested that, using this effect, a clock could be built that relied on the absolute measure of physical phenomena rather than somewhat arbitrary and changeable astronomical measurements. By the time Essen started his work, there had already been various attempts at making an atomic clock in the United States using the natural frequencies of compounds such as ammonia, but all had proven disappointing. Essen, however, thought that perhaps combining Rabi's method with his quartz-ring oscillator might do the trick. In 1955, he achieved his goal. Essen was so excited he dragged his boss at the National Physical Laboratory into the lab "to witness the birth of atomic time." (The "atomic" in "atomic clock," incidentally, refers to the properties of atoms, not atomic fission as in a nuclear reactor or nuclear bomb.)[24]

The atomic clock works something like a microwave oven, which uses electromagnetic energy to excite the water molecules in food and heat them up. However, instead of reheating frozen burritos, an atomic clock works by using energy to excite cesium, a highly reactive elemental metal with a very low melting point. Though it is difficult to handle, Essen chose to use cesium-133 as his material because, as one of the alkaline metals (group 1 on the periodic chart), it only has one electron in its outermost valence shell. The least electronegative of all stable elements, cesium is rather willing to give up that loosely held electron, which means, for purposes of magnetic resonance, it gives a very clear and strong signal. (The alkaline metal francium is below cesium on the periodic chart, but, besides the fact that cesium is even more electronegative, francium is also highly unstable and radioactive. In fact, most isotopes of cesium are also radioactive, but cesium-133 is not. Some atomic clocks use hydrogen or rubidium, which are above cesium in group 1, but these are slightly less accurate.)

Another important difference is that, whereas a microwave oven operates between 2,450,000,000 hertz and 915,000 hertz, atomic clocks operate at a frequency of precisely 9,192,631,770 hertz (that is, about 9.19 gigahertz). This is the optimal frequency to excite the valence electrons of a beam of heated cesium atoms, which then trigger a detector. This detector provides feedback to the quartz timekeeping element, which is set up so that feedback keeps it in synchronization (resonance) with the vibrations of cesium: if the frequency slips off-target, the feedback instantly corrects it. Keeping time is thus a matter of counting the cycles per second and dividing by 9,192,631,770. In other words, the excited cesium atoms act like the pendulum in a longcase clock: as a harmonic oscillator device to regulate the mechanism.

It was quickly realized that the time kept by an atomic clock is far superior to ephemeris time (though, of course, ephemeris time was used to calibrate atomic clocks). In fact, the cesium clock is the most accurate instrument that humanity has ever invented: Essen's first atomic clock was accurate to one second in 300 years, and more recent cesium atomic clocks, which use lasers to cool the cesium down to near absolute zero, are accurate to within an astounding one second in 1.4 million years. This is much more accurate than using the rotation of the earth. In fact, defining the day as 86,400 UTC seconds on a slowing earth means that midnight will eventually move toward solar midday, so we sometimes have to add leap seconds to keep atomic clocks in tune with nature.

What's more, atomic clocks are versatile: different designs using the same principle can be made as small as a computer chip and sent into deep space. The short-lived ephemeris standard (by which the cesium clock was calibrated) was replaced in 1967 when the General Conference on Weights and Measures linked the second to the atomic standard; as the official brochure puts it, "The second is the duration of 9,192,631,770 periods of the radiation corresponding to the transition between the two hyperfine levels of the ground state of the cesium-133 atom." Atomic time is transmitted to the public by such sources as the National Institute of Standards and Technology's radio signal, which broadcasts at 60 kilohertz out of Boulder, Colorado, and the German DCF77 transmitter. (A whole classification of timekeeping devices, *radio clocks*, work off such transmitters. Some companies market them as "atomic clocks," but they're really a sort of primary/secondary clock.)[25]

Our certainty of the measurement of time is now so accurate is that other units of measure are linked to the second. For instance, in 1983, the meter was redefined as the distance light travels in 1/299,792,458 of a second. The newton is the force needed to accelerate one kilogram

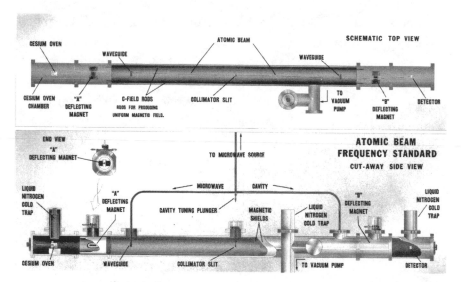

Figure 5.2. Atomic clock diagram. National Institute of Standards and Technology, courtesy Wikimedia Commons

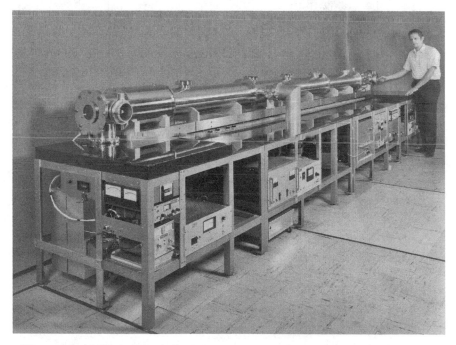

Figure 5.3. 1950s atomic clock. National Institute of Standards and Technology, courtesy Wikimedia Commons

of mass at the rate of one meter per second squared in the direction of the applied force. After the 2019 redefinition by the International Bureau of Weights and Measures of the seven basic units in the International System of Units, only the mole (a quantity of elementary particles) is not in some way defined by the second. If Einstein returned us to a relativistic world, the atomic clock turned Aristotle's logic, where time is knowable only by the universe of moving things, on its head: the universe of moving things is knowable because we have a precise and unchanging measure of the second.

Time in Information Technology

While atomic clocks keep precise scientific time, our day-to-day world is controlled by computers, and the time as shown by computers has replaced the bell tower or the railroad station's clock as the primary indicator of social time. Computer time is at once individual—as displayed by our smartphones—and communal—airline schedules, class times, and electronic systems. In this last section, we will look at how computers keep time and why, in the new millennium, it was widely feared that a failure in this system would disrupt life as we know it.

All computers have internal electronic clocks—really, just a simple oscillator. However, it's important to know that computers use two different sorts of "time": *wall clock time*, which is time as perceived by human beings, and *process time*, which measures the work done by the central processing unit (CPU) and does not include programmed delays such as waiting for input. All CPUs are rated in hertz; the more hertz, the faster they operate—and the more instructions they can execute at once. Of course, the two must necessarily work well together, such as in a program that coordinates an airline's schedule.

A computer's clock calculates time—ticks—from a certain arbitrary epoch. Unix systems, for instance, calculate time from midnight on January 1, 1970, while Windows uses the same date from 1601. The electronic oscillator triggers a timer chip that operates at a certain frequency (anywhere from 18.2 to 1024 hertz) and periodically "ticks"—adds 1 to the count from which the epoch began. For human use, then, this is converted to usual dates: 1,474,569,180 seconds since the beginning of the Unix epoch is 6:33 PM UTC on September 22, 2016. A computer clock is not limited to seconds, of course: it can count time in arbitrarily fine units, limited only by storage capacity. This is called *resolution*.[26]

Computer clocks are not necessarily very precise—able to keep time to within a fine resolution—or accurate—that is, faithful to a universally accepted time standard such as UTC. There is also a certain amount of random "jitter" when taking readings. This is why they automatically correct themselves based on an external time signal. Mobile phone clocks, for instance, have digital clocks like computers do, but they also adjust themselves according to what the network tells them. Still, for day-to-day use, we don't think very much about computer timekeeping—we just use it. It was this blind reliance that led to "the disaster that didn't happen": the year 2000 bug, or Y2K for short.

The Y2K problem

As the twentieth century drew to a close, many computer experts warned that, as the chronographic odometer ticked over from December 31, 1999, to January 1, 2000, computers would think it was January 1, 1900, and the software would fail catastrophically. Systems that automatically sent out social security checks would think that the recipients had not even been born yet, power grids and assembly lines would shut down, people would be charged decades of interest on loans, trains would collide, the GPS system would malfunction, nuclear missiles would lock out those trying to access them, and all sorts of chaos would ensue. This was not due to any human incompetence but rather to the cleverness of programming pioneers: early computers had limited storage capacity, and memory cost as much as a dollar per bit, so they saved space by recording the date as two digits, rather than four—"75" instead of "1975." When much of this software was written, most work was still done manually; however, by the year 2000, computers controlled everything from aircraft to toaster ovens—often using software that no one would have thought would still be in service when it had been written 30 years earlier. (A secondary problem: some programmers were not aware of the Gregorian reform calendar rule and didn't know that 2000 would be a leap year.)

The history of Y2K is yet to be written, in part because we don't know the true cost or how close we came to disaster—understandably so, as organizations sought to cover up the worst of the glitches. Though experts such as the IBM computer engineer Robert Bemer (1920–2004) had been warning of the impending problem since at least the 1950s, both the government and private industry continued to use two-digit dates—following the lead of the Department of Defense, which insisted on the practice as a cost-saving measure.[27] The problem was compounded by late modern capitalism's unwillingness to plan for the long term or admit the fragility of systems where it might upset profits. All of this changed in the last years of the twentieth century, when the Y2K bug quickly took on the dimensions of a modern millennial scare—or, as the cover of the January 1999 issue of *Time* magazine demanded of the reader, "The End of the World!?!"

The hysteria did serve a useful purpose, however: it persuaded leaders that there was, indeed, a problem. The result was that thousands upon thousands of programmers had to be hired to go into perhaps billions of lines of code written in a plethora of programming languages, change it to deal with four-digit dates, and then find every instance of a two-digit date and change it to a four-digit date. (Other fixes were, for large databases, to change six-digit month-year-date formats to a three-digit year and a three-digit ordinal day in the year; or to write in workarounds to calculate correct values from two-digit dates). This required millions of hours of labor and untold expense. A particular commercial business—say an airline—might work using thousands of interconnected machines from web servers to reservation software to flight computers, all of which share data, and some of which might be Y2K compliant while the rest were giving out bad data. Some panicked analysts predicted the final costs would add up to $3.6 trillion; the actual cost was $300 billion ($450 billion in 2020 dollars).[28]

In the end, there were fairly few bad malfunctions: some US government reports had the wrong date on them, a few spy satellites sent back garbled data for several days, about 10,000 credit card machines didn't work, a few bank payments and money transfers were delayed, drivers in Indiana were issued licenses that expired after five years instead of four, a customer in New York was charged $91,250 for renting a videocassette for 100 years, a number of Microsoft programs became obsolete, and both the official French weather forecasting service and the US Naval Observatory's clock decided the date was 1 Jan 19100. The worst thing that happened was that 154 pregnant women in Sheffield, England, were given incorrect Down syndrome test results, with two healthy babies being aborted and four babies with the disorder being born whose mothers had been told they were at low risk.[29]

Whether this limited effect was due to proactive vigilance on the part of governments and private industry, or whether the world was too complex a system to come crashing to an end because of some buggy code, we may never know. South Korea and Italy—as well as most US school systems—did very little to prepare for the changeover yet had few or no difficulties.[30] Nonetheless, the Y2K problem highlights one important fact: knowing the accurate time, in the modern world, is more than orienting ourselves or coordinating society. Thanks to our reliance on information technology, it is critical to the economy and, on a fundamental level, to the operation of the world itself.

The apocalyptic thinking surrounding the Y2K problem shows that certain aspects of timekeeping have remained constant from the earliest eras. This sort of concern is hardly uncommon; many religious sects throughout history have believed that the "end of time" would come, bringing with it judgment and an overturning of the order of the world. (We call the fear that the end of cosmic cycles of time will lead to a new era for humanity *millennialism* after the "thousand-year kingdom of the saints" spoken of in the book of Revelation.) What Y2K did was take this ancient concept, secularize it, and place it in machines and systems whose mechanisms are as mysterious to most modern people as the workings of the heavens were to premodern peasants. The fear of the new millennium shows that, no matter how sophisticated our technology, we still regard the art of timekeeping with a sort of religious awe.

In Conclusion . . .

The twentieth and twenty-first centuries not only introduced an unprecedented accuracy and precision to timekeeping; they made this technology ubiquitous and put it in the hands of anyone able to buy a mobile phone or cheap watch. In the past, scholars studied the movements of the heavens to gain insight into the universe. Certainty of the time and season was foremost among these spheres of knowledge; by the careful study of the sun, moon, stars, and planets, they could determine the unchanging laws of the universe. Although Western culture was not unique in the special honors it gave to astronomy, it was unique in the emphasis it placed on precise measurement and its willingness to both undergo profound changes from the results of this study—the Copernican revolution, Newtonianism, industrialism, the digital revolution—and to put this knowledge in the hands of ordinary people. This had profound mental and social implications, giving rise to social coordination, an internalized sense of time, life-changing technology, and a reorganization of society of itself. The study of the times of the heavens led not only to modern science but also to the very mentality of modernity.

In some ways, because of the theory of relativity, our own era echoes the thinking of medieval scholars who, following Aristotle, held that the existence of time depended on the universe of moving things. In others, because of the seemingly sure, scientific, and universal knowledge of time, it fulfills Newton's ideas of absolute time. The atomic clock is the most accurate and precise machine humanity has created—so much so that, rather than time being knowable only by moving things, today, all things are known by time. We now know the movements of the heavens, seemingly so eternal to the ancients, are now changeable; only the infinitesimally small—indeed, the atomic—is constant. In fact, it is the heavens that are now measured according to the clock, rather than the other way around. The machine itself, not the observation of some external object, is what defines reality.

How strange this would have seemed to a medieval scholar: while society itself potentially becoming disordered for not knowing the proper time would certainly seem to be within the realm of possibility, the idea that that we couldn't simply look at the stars to reset our time-keeping devices, but rather had to first rely on our machines, would have been a bizarre concept! Yet, despite this complete overturning of the logic of timekeeping from the primacy of the natural world to the human-created machine, the importance of timekeeping itself has remained consistent: all other knowledge, from the most mundane output of computer programs to the measurement of the distance of the *Voyager* probes from Earth to experiments that push the boundaries of human knowledge, depends on the readings of extraordinarily accurate clocks. Knowledge of the right time, in other words, is still the science upon which all other activities depend.

The following exercises are intended as practical, hands-on adjuncts to the rather abstract information presented in this book. I am a great proponent of active learning, and by working your way through these experiments and exercises, you can internalize your understanding of the development of various timekeeping concepts.

Chapter 1 Exercise: Constructing an Analemmatic Sundial

Marcus Vitruvius Pollio, most commonly called simply Vitruvius, was a Roman architect who wrote his treatise *On Architecture (De architectura)* in about 25 BCE.* Like most Romans, he was more concerned with practical results than with theory. Unlike Ptolemy's highly theoretical treatise, Vitruvius gives us just enough astronomy to tell us how to construct a sundial. Nonetheless, Vitruvius's treatise was an important means of transmitting the knowledge of the ancient world to the Middle Ages. This is an abbreviated version of his instructions for constructing a sundial taken from Evans's excellent *History and Practice of Ancient Astronomy*. Doing this exercise is an excellent way to really understand ancient astronomy through active learning and will show you just how clever the ancients were in using mathematics and geometry to create precise rules for making observations.

Making a sundial is complicated by the fact that the sun's path through the sky changes over the year, which means that the length of the shadows cast by the *gnomon*, the shadow-casting part of the sundial, will vary as well. As far back as Hipparchus, the ancients used a clever geometrical device called an *analemma*, which is the Greek word for the pedestal of a sundial, to show how the shadows cast by the sun change over the course of the year. (Note that this analemma is different from the analemma shown on a globe, which symbolizes the fact that the earth's orbit is not a perfect circle, and so the position of the sun as photographed at noon—or any other fixed time—will vary during the tropical year along a path that looks like a figure eight. I discussed this analemma in chapter 4.)

Vitruvius gives his instructions on how to construct an analemma in the ninth book of *On Architecture*. Essentially, what we will be doing is drawing a model of the universe seen from the side, and then projecting this onto a top-down view of our sundial pedestal. To do this exercise, you'll need a compass, a straightedge, and a protractor.

First, draw a circle, the *meridian circle*, to represent the course of the sun through the sky on the equinoxes. The center of the circle is point A. Point A will also be the tip of our gnomon, which I have drawn as a simple gray line. Draw a ground line for the gnomon to rest on, and, parallel to that, running through A, draw line EI bisecting the meridian circle to represent the horizon. At the base of the gnomon, mark point B. Now, draw line NF at an angle to the ground line, θ, equal to the latitude of the place for which you are constructing your sundial. This represents the axis of the universe. Where line NF intersects the ground line, make point C; line BC is the length of the shadow of the gnomon on the equinox. (See figure A1.1. Vitruvius actually has us take the proportions of the shadow to the gnomon first, but it is easier for us to simply measure with a protractor. I have made θ equal to 42°, to match the approxi-

*Adapted from Evans, *History and Practice of Ancient Astronomy*, 135–139.

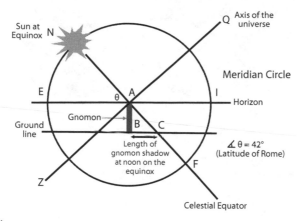

Figure A1.1

mate latitude of Rome.) Finally, draw line QZ at a 90° angle to NF to represent the axis of the universe.

So much for the equinox; what about the solstices? To find out where the sun is at these times, we'll need to draw two more circles (or, rather, semicircles). To do this, Vitruvius has us mark off 1/15 of the circle to either side of point N to show us where the sun will be at the summer and winter solstices, which we can do by simply measuring 24° (1/15 of a circle) to either side of line NF with a protractor and drawing lines through point A up to the meridian circle and down to the ground to show us where the sun will be and where the gnomon's shadow will fall at noon on those dates. These will be lines MG and LH. (See figure A1.2. Vitruvius would have probably found the points first by subdividing

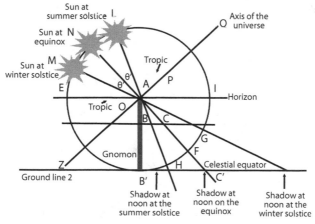

Figure A1.2.

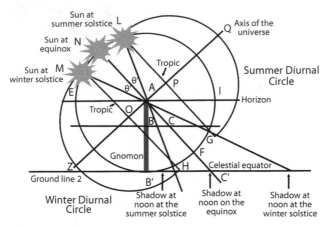

Figure A1.3.

the circle and then drawing the lines, but, for those not trained in practical Euclidian geometry, the protractor is easier. Note that I have marked the 24° angles as θ′ in the diagram. I have also moved the ground line for clarity, but the angles and proportions remain the same. Points B and C transposed to our new ground line are marked as B′ and C′.)

Now, it is time to draw the *diurnal circles* to show us the sun's path at the solstices. Draw lines LG and MH parallel to NF to represent the tropics. Where these intersect QZ, the axis of the universe, are points O and P. Place your compass on points O and P and draw the semicircles that represent the sun's path on the solstices (see figure A1.3). Now find the place where the horizon

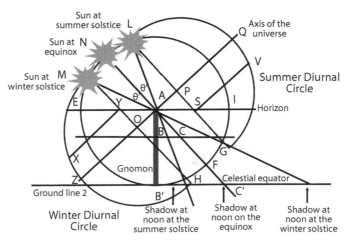

Figure A1.4.

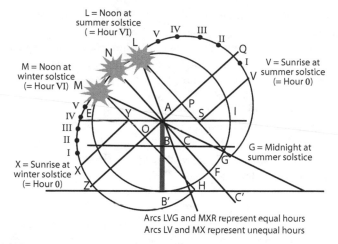

Figure A1.5.

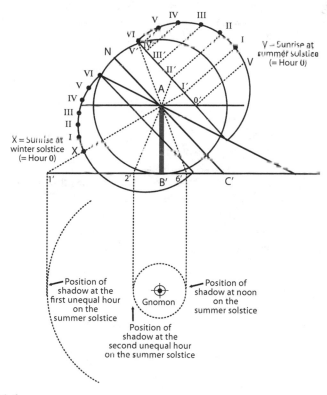

Figure A1.6.

line EI intersects the two tropic lines, LG and MH. Mark these points as S and Y, respectively, and draw two lines from these points parallel to the axis of the universe until they hit diurnal circles. Mark the intersection points V and X, respectively; they show us where the sun will rise on the solstices—or, if you prefer, how much of the sun's circular path will be above the horizon on those days (see figure A1.4).

We are finally ready to draw the shadow tracks on the base of the sundial. We do this by dividing the diurnal circles, but we have some choices to make here. The first is numbering: The Romans, who divided the day and night into twelve unequal hours each, would have marked noon as *sextus*, or six, so we'll do the same, even if we're used to noon being 12:00. The second is whether we wish to use equal or unequal hours; if equal, we divide all of arc LVG into equal sections, if unequal, we divide just the daylight hours, which is arc LV. To keep with the Roman way of doing things, I have chosen to divide arcs LV and XM into six increments, marked with Roman numerals from I to V (letting L and M stand for VI), but remember these represent not modern clock time but rather the *unequal* hours (see figure A1.5). You can see how the winter hours are smaller than the summer hours.

The next step is to make a top-down diagram to show the shadow track. Directly under point A, but some ways down the sheet of paper, draw the top of your gnomon as if you are looking down at it. Draw lines parallel from the hours marked on the diurnal circles to the tropic. I have used dotted lines for clarity marked the intersection points I′ to VI′. Now draw lines from these places through point A (the tip of the gnomon) down to the ground. I have again used dotted lines marked these intersections with Arabic equivalents of the Roman numerals—1′, 2′, and 6′. (For clarity, I will show only the shadow tracks for three representative hours.) Note that the early morning ones will be *behind* the gnomon. Open your dividers to the length of the shadow (i.e., the space from point B to 4′, 6′, etc.). Draw circles of this radius around the mark you made to represent the top of the gnomon. Now, draw lines straight down from those points to the circles you just drew; the point where they intersect is where the top of the shadow will be during that (unequal) hour. (Again, I used dotted lines.) To do this for the winter solstice, merely repeat the process for the other diurnal circle (see figure A1.6). Finally, you can do the experiment again using equal hours.

Chapter 2 Exercise: Finding the Equal Hour with an Astrolabe

Copy the astrolabe images on pages 187–193 with a photocopier and assemble them with a paper fastener through their center. It's best to copy the rete onto a transparency, so you can see the front through it. Be neat; your astrolabe's precision depends on getting the hole as close to the center as possible. This astrolabe is calculated for 49° north latitude (the latitude of Paris), but you can create one for your own location at http://astrolabeproject.com.

Astrolabes can, of course, be used for a variety of calculations, but we're going to concentrate on using our simplified version to find out the time. First, you need to find out the *astrological date* to correct the calendar day to where the sun is in its annual journey around the ecliptic. Suppose that today is September 26. Looking at the back of the astrolabe, we find that this corresponds to Libra 3.

Now turn the astrolabe over. Note that the zodiac is also shown in a circle on the rete, while the mater has a series of concentric circles marked on it. Each represents a five-degree increment of elevation in the sky. The closer to the center of the astrolabe a line is, the higher an elevation it represents—the line at the top is 20°, while the center of the circle is 90° (directly overhead). The lines are basically a stereoscopic projection of the sky at the latitude for which this particular plate is calculated—in other words, an analemma. It uses trigonometric functions to predict when certain objects will be at certain places in the sky. We already saw an example of how to calculate this using simple geometry with our Vitruvian sundial, but Arabic astronomers worked out how to do this mathematically for a number of celestial objects.

You'll notice that the zodiac circle on the rete is offset from the center. If you move the rete around, you'll see that, just as the sun doesn't climb very high in the sky in the winter months, the zodiac signs around the time of the winter solstice—Capricorn and Sagittarius—get all the way up to around the 20° mark on the mater. Cancer, which has the summer solstice, stays down by the higher elevations. In other words, the zodiac circle mirrors the sun's path in the sky.

Look at the left side of the astrolabe. Let's suppose that on the afternoon of September 26 in Paris, we sighted the sun at an elevation of 40°. Move Libra 3 to the 40° on the right side of the astrolabe. Use the alidade to read directly to the numbers on the outside. You'll see it's about 12:45 PM.

You can use the same procedure with any of the stars on the rete (provided you know how to find them in the sky) to find out the time at night. You can also figure out at what time the sun will rise and set, since the outermost, darkest circle represents 0° of elevation. If today is Libra 3, we can read that the sun will set a little before 6 PM. Astrolabes can be used to render celestial observations to both equal and unequal hours; however, for the sake of simplicity, ours calculates only equal hours.

Astrolabe images courtesy of Richard Wymarc.

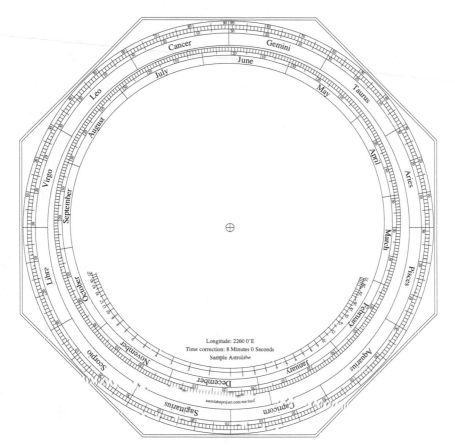

Longitude: 2260 0'E
Time correction: 8 Minutes 0 Seconds
Sample Astrolabe

astrolabeproject.com me tool

Figure A2.1. Astrolabe mater

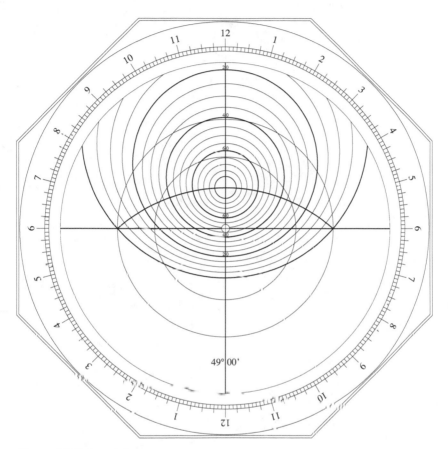

Figure A2.2. Astrolabe tympan

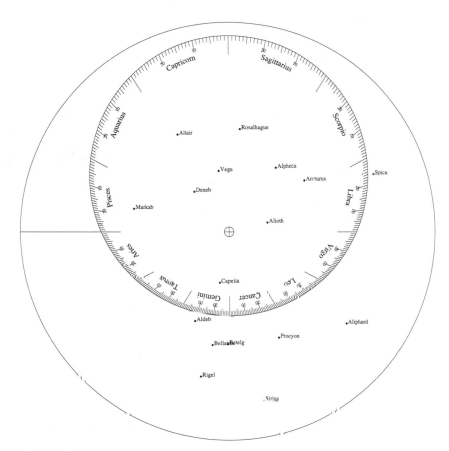

Figure A2.3. Astrolabe rete

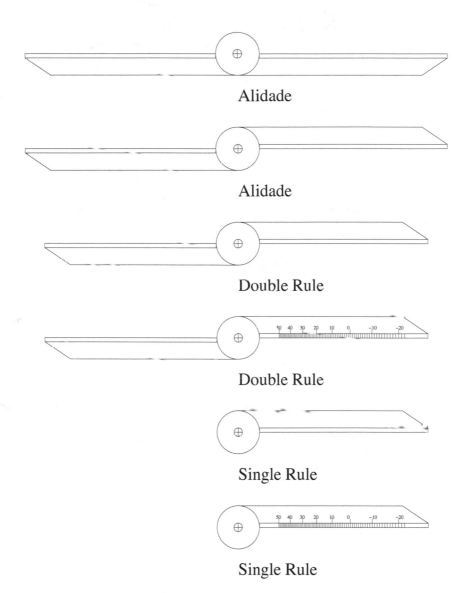

Figure A2.4. Astrolabe rules

Chapter 3 Exercise: Pendulum Isochronism

For this experiment, you will need a length of string; something overhead to tie it to, such as a light fixture, overhead beam, or ceiling-mounted projector; and some different objects of known weight such as golf balls, fishing weights, or fencing foil and épée pommels; a protractor; and the stopwatch function on your smartphone. (Use common sense and don't tie your string to anything delicate or breakable, like a sprinkler head.)

On a piece of notepaper, draw out a grid to note the time of the swings, changing one variable for each grid. For instance:

Golf ball (46 g) with 50 cm of string at 60° of arc

Foil pommel (116 g) with 50 cm of string at 60° of arc

Épée pommel (150 g) with 50 cm of string at 60° of arc

Épée pommel (150 g) with 1 m of string at 60° of arc

Épée pommel (150 g) with 25 cm of string at 60° of arc

Épée pommel (150 g) with 50 cm of string at 30° of arc

Épée pommel (150 g) with 50 cm of string at 90° of arc

Measure each swing with your stopwatch. What variations do you note?

Consider how you would make these observations if you were not able to time the pendulum's swings in minutes or seconds but instead had to use a water timer, as Galileo did. You can approximate a water timer with a common sink faucet, a water cooler, or a large container of bottled water with a spigot.

To catch the water, use a graded beaker or even a regular plastic drinking cup, and to measure the amount of liquid, you can weigh it, as Galileo did, use a common kitchen measuring cup, or, less accurately, simply make marks on your cup at one-centimeter intervals—though this will work best if your cup is a cylinder.

Try timing the pendulum by turning the spigot or sink on full-force and comparing the amount of water produced when you change each variable. This experiment should give you some idea of the difficulty of doing precise scientific experiments before the invention of timekeeping devices with second hands. In fact, the famous historian of science Alexandre Koyré held that Galileo's experiments timing balls rolling down inclined planes would have been impossible to perform, but in 1961 Thomas B. Settle, then a graduate student at Cornell University, succeeded in reproducing them.*

---

*"An Experiment in the History of Science," *Science* 133, no. 3445 (1961): 19–23.

Chapter 4 Exercise: Finding Position with a Chronometer

Because lines of longitude converge (run together) the farther north you go, the distance between them shrinks in accordance with a cosine function. The formula to convert one degree of longitude to units of distance is the cosine of your latitude times the length of the degree at the equator (69.17 miles or 111.32 kilometers), with the degrees of latitude expressed as a decimal. Degrees are made up of 60 minutes, so 2 degrees 15 minutes would be expressed as 2.25, because 15 minutes is 1/4 of a degree. (For more precise location, each 60-minute degree also has 60 seconds, which are usually noted by a decimal fraction.)

The precise formula (assuming the earth is a sphere) is:

$$\Delta\ 1°\ \text{Longitude} = \frac{\pi}{180}\ a\cos\Phi$$

where a is the radius of the planet and Φ is the longitude in degrees. (On Earth, $\pi/180\ a$ works out to be the aforesaid 69.17 miles or 111.32 kilometers.) In other words:

At the equator (0°), a degree is 69.17 miles (111.32 km)
At 15°, a degree is 66.83 miles (107.55 km)
At 30°, a degree is 60.0 miles (96.49 km)
At 45°, a degree is 49.0 miles (78.85 km)
At 60°, a degree is 34.67 miles (55.8 km)
At 75°, a degree is 18.0 miles (28.9 km)
At the North Pole (90°), where all the lines of longitude converge, there is no distance between them.

Given these figures, answer the following questions:
Example 1: You are sailing at the equator. Local time is 12:00 hours (noon). Greenwich mean time is 14:00 hours (2 PM). Are you east or west of the prime meridian, and how far in miles and kilometers? Consulting a world map, what is your nearest landmass?

Example 2: You are at 15° north latitude. Local time is 18:00 hours (6 PM). GMT is 22:00 hours (10 PM). Are you east or west of the prime meridian, and how far in miles and kilometers? Consulting a world map, what are you near?

Example 3: You are at 15° north latitude. Local time is 09:00 hours (9 AM). GMT is 01:00 hours (1 AM). Consulting a world map, what city are you near?

Example 4: Continuing from example 3, a weather station reports a typhoon heading due west toward your position at 30 knots (one knot is 1.15 mph, or 1.85 kph). The weather station's chronometer is broken, but the local time is a half-hour later than yours. How long before the typhoon hits the city?

Example 5: St. Petersburg, Russia, is located at almost exactly 60° north and 30° east of Greenwich. Suppose that people in St. Petersburg see something unusual in the sky at 11:48 PM: it's a large meteor burning up in the earth's atmosphere! People in a small town the middle of Siberia 1,674 kilometers due east of St. Petersburg report a fireball directly overhead at 01:50 AM local time. How fast was the meteor going in terms of ground speed?

Chapter 5 Exercise: Tolman's Paradox and the Impossibility of Time Travel

The math around relativistic physics isn't always complicated: we can see the truth about some theories with simple algebra. For instance, here we'll disprove the possibility of faster-than-light travel—and look at time travel, as well.

Relativity states that no signal can travel faster than light. If there was no absolute speed limit, if we *could* travel faster than light, we could relay information into the past. This leads to something called *Tolman's paradox*, after the American physicist Richard Chace Tolman (1881–1948).

Suppose that two spaceships—the *Enterprise* and the *Discovery*—pass one another moving at a relative speed of .8 *c*, 8/10 the speed of light. (Of course, the crew of *Enterprise* would see *Discovery* as moving at .8 *c* and themselves as stationary, and vice versa.) Let's now suppose that both spaceships have a hyperspace radio transmitter—a "tachyonic antitelephone"—capable of sending information at 2.4 times the speed of light as measured from each ship. Three hundred days after they pass, the *Enterprise* has a reactor leak. Its crew, dying of radiation sickness, sends a last message to *Discovery*: "Remind us to check our reactor shielding."

Now, remember, from *Enterprise*'s perspective, *Discovery* has been moving away at .8 *c* for 300 days. The message is therefore sent when *Discovery* is 240 light-days away, and it will catch up when, from *Enterprise*'s frame, *Discovery* is 450 light-days away. However, because of time dilation, the *Discovery* thinks that only 270 days have passed since the message was sent! We can get this from the time-contraction formula:

$$\frac{1}{\tau} = \sqrt{1 - v^2 / c^2}$$

Time proceeds for *Discovery* at 0.6 the rate it does for *Enterprise*. Now, let's suppose that *Discovery* instantaneously sends the reply "Check your reactor shielding!" From the perspective of the crew of *Discovery*, *Enterprise* is 216 light-days away (270 days times 0.8 times the speed of light). From *Discovery*'s perspective, the message will reach *Enterprise* on day 270 (the time in their own frame when they sent the message) + 135 days (transit time), or on day 405. But because time dilation is symmetrical, *Enterprise* receives the reply on (their) day 243: 57 days before they sent the original message! The reactor leak that prompted the *Enterprise*'s first message would have paradoxically never occurred!

Absolute time: The belief that time is a real thing, proceeding at its own pace independent of any other object.

Ab Urbe Condita (AUC): A Roman dating system. Latin "from the founding of the city [of Rome]," which supposedly took place in 753 BCE.

AD: Abbreviation for *anno Domini* (year of the Lord), first introduced by the English monk Bede (c. 672–735 CE). A system of dating that measures from the conception of Jesus as calculated by the sixth-century CE monk Dionysius Exiguus. Today abbreviated CE.

Almagest: A book by the ancient writer Claudius Ptolemy that remained the standard astronomical textbook until Copernicus developed his heliocentric model of the universe in the sixteenth century CE.

AM: *Ante meridiem,* "before noon"—*meridies* being the Latin for "noon."

Analemma: (1) The Greek word for the pedestal of a sundial. (2) A geometrical method to show how the shadows cast by the sun change over the course of the year. (3) A figure-eight diagram showing how the sun's position at noon varies during the *tropical year*.

Analog computer: A device that uses a continual physical phenomenon, be it mechanical, hydraulic, or electric, to record data; unlike a digital computer, it is not programmable.

Anchor escapement: A type of *escapement* invented by Robert Hooke in the seventeenth century. It consists of a gear with pointed teeth, called an escape wheel, and an anchor mounted just above it on a pivot. The teeth of the escape wheel interact with the curved parts of the anchor, which are called pallets. The pivoting of the anchor, in turn, is driven by the clock's pendulum. The device is made so that, as the anchor pivots back and forth with the swinging of the pendulum, the pallets alternately catch and release the teeth of the escape wheel. This drives the escape wheel, which in turn drives the clock gearing at a pace regulated by the pendulum.

Anno Hegirae (AH): The Islamic year, dating from Mohammad's flight from Mecca to Medina in 622 CE.

Antisynchronization: A property of pendulums attached to the same beam; one will come to swing opposite the other.

Apparent solar time: The local time as determined by observation, say, by using a sundial or a sextant. Contrast with *mean solar time.*

Aristotle: Ancient Greek philosopher whose maxim "time is the number of motion with respect to the before and after" was immensely influential on the development of premodern science.

Astrarium: An *astronomical clock* built by Jacobo Dondi in the early fourteenth century.

Astrolabe: A form of analog computer used to make astronomical observations.

Astronomical clock: A clock that models various astronomical phenomena in addition to the primary ones by which time is told.

Atomic clock: An *electronic clock* that uses the known frequency of atoms changing energy states as its *oscillator*.

Azimuth: In astronomy, the plane of the horizon relative to the observer.

B'ak'tun: The Mayan name for a period in the Mesoamerican calendar that lasts about 394 1/2 years.

Balance spring: An *isochronic* timing device that works in conjunction with a *balance wheel* to regulate the movement of a small clock or watch's mechanism. Every time the *balance wheel* is moved forward by the clock's mechanism, it is pushed back by the balance spring. The combination of the inertia of the wheel and the reliable resistance of the spring gives a very constant result.

Balance wheel: An *isochronic* timing device that works in conjunction with a *balance spring* to regulate the movement of a small clock or watch's mechanism. Every time the balance wheel is moved forward by the clock's mechanism, it is pushed back by the *balance spring*. The combination of the inertia of the wheel and the reliable resistance of the spring gives a very constant result.

Bar balance: An *isochronic* regulator invented by John Harrison consisting of two weighted bars connected by balance springs.

Bells: One type of public time signal, used extensively in medieval Europe. In the *Explanation of the Divine Offices* (*Rationale divinorum officiorum*), a handbook of church furnishings by the French bishop Guillaume Durand written sometime before 1286, there are six types of bells used in a church to ring the *canonical hours*: the squilla (hand bell), the cymbalum (in the refectory), the nola (in the cloister), the nolula or double campana (in the clock), the signum (in the tower), and the campana (in the bell tower).

Butterfly escapement: See *Hipp toggle*.

Calendar: A system for organizing days for social and administrative purposes, usually based on natural cycles.

Calendar Round: A period of 18,980 days, or about 52 years, in the Mesoamerican calendar.

Cam: An eccentrically shaped gear.

Canonical hours: The eight prayer times of Christian monks: Matins (sunrise), Sext (midday), Compline (sunset), and Laudes (around midnight), to which were added the Roman quarter hours mentioned in the Bible, whose timing depended on the natural signals: Prime (shortly after Matins); Tierce (later in the morning), None (which was originally midafternoon, but gradually moved closer to modern noon, to which it gives its name, over the course of the thirteenth century), and Vespers (around sunset).

CE: The "Common Era"; the modern secular rendering of AD.

Celestial sphere: A coordinate system used in astronomy, projecting the earth's system of latitude and longitude onto the apparent sphere of stars surrounding the earth.

Chinese calendar: A lunisolar calendar, maintaining separate lunar and solar cycles. The first of the lunar month is defined as the new moon, while the solar calendar breaks the year down into months of 30 or 31 days.

Chronometer: A timepiece accurate enough to determine longitude when compared to local time.

Clepsydra: Greek for a *water clock*.

Clock: A machine that takes a continuous physical phenomenon that happens at a known rate—be it gravity pulling on a weight or causing water to flow from a reservoir into a receptacle, the unwinding of a spring, or electronic impulses—and makes how much of that phenomenon has occurred readable to humans—whether by a measure of liquid, hands on a clock face, or a digital display—so that an observer can know how much time has passed.

Computus: The rules for determining the date of Easter.

Cross-beat escapement: Jost Bürgi's late sixteenth-century invention to improve the accuracy of a *spring-driven clock*, consisting of two foliots (see *virge and foliot*) traveling in opposite directions in order to equalize the forces acting on the clock. The earliest surviving example of this is the calendar clock he built for William IV of Hesse-Kassel, a famous patron of scientists. This clock kept its accuracy for a remarkable three months between windings. His work also included a *remontoire*.

Cycloid: The sort of curve a point on the rim of a wheel would trace out as it rolls along. Cycloid-shaped pendulum suspensions were used by Huygens to improve the accuracy of his clocks.

Daylight savings: The practice of turning clocks forward in the spring and back in the fall to reduce energy consumption and gain more hours of daylight during the winter months.

Deadbeat escapement: Invented by the English mathematician and astronomer Richard Towneley in the 1670s and manufactured by the clockmaker Thomas Tompion, this improves upon the *anchor escapement* by eliminating the recoil. Instead, the pallets catch on the escape wheel, preventing movement. This eliminates the inaccuracies caused by recoil and is still used in some modern pendulum clocks.

Decimal time: Measuring time in base-10, instead of base-12.

Declination: The equivalent to latitude on the *celestial sphere*—the distance above or below the celestial equator, measured at a 90-degree angle.

Duodecimal: Base-12.

Durational time: See *relative time.*

Ecliptic: The path of the sun's apparent motion against the backdrop of the stars over the course of the year (caused by the fact the earth is traveling around the sun). The circle of the constellations chosen to form the zodiac, not coincidentally, follows the ecliptic—which is also the plane of the planets' orbits around the sun—since it is the backdrop against which ancient people made most of their observations.

Electric clock: A clock that is powered in part or whole by electricity. The most traditional are mechanical clocks that simply use electricity to wind the spring, raise the weight, push the pendulum, or use an electric-powered *remontoire*. In these cases, the timekeeping mechanism is still a balance spring, pendulum, or other mechanical device. If, however, the clock's timekeeping mechanism is powered by electricity, it is called an *impulse clock*; if electricity is used only to rewind the clock, it is called a *rewound clock*.

One of the first was invented by Carl August von Steinheil (1801–1870), a professor at the University of Munich, who in 1839 synchronized two distant clocks by placing a rocker switch under the pendulum of a regular mechanical clock. Every time the pendulum swung, it tripped the switch and sent an impulse through a telegraph wire over a distance of two kilometers to a solenoid (a coil of wire that acts as an electromagnet when a current is sent through it), which then acted on a permanent magnet affixed to the anchor escapement of another clock. The secondary clock was thus automatically synchronized to the main one.

The *tuning-fork clock* was first described by 1866 by Alfred Niaudet-Bréguet (1835–1883), who came from a family of clockmakers and whose grandfather had developed a master clock that synchronized and wound pocket watches. Because tuning forks vibrate at known frequencies, they can be used for precise time measurements. Niaudet-Bréguet created an impulse clock that used the frequency of an electrically driven tuning fork to control a small escape wheel; Hippolyte Fizeau used a precise clock built on this principle in his experiments measuring the speed of light. Almost a century later, Max Hetzel of the Bulova Watch Company employed same idea in the company's groundbreaking Accutron watches.

The rise of utility companies meant that it was possible for electricity to replace mainsprings on a widespread basis. *Synchronous clocks* rely on the power grid itself, using the frequency of the alternating current to keep time. (Alternating current reverses direction many times per second; direct current is a steady flow. Alternating current is a more efficient means of transmitting electricity, which is why it is used by utility companies.) Synchronous clocks have no oscillator; the electricity flowing through the electrical grid is itself the timekeeping element. Because they rely on an external factor, these clocks are very accurate—unless there's a blackout. They were invented by Henry Ellis Warren (1872–1957) in 1918 and gained in popularity in the 1930s. In essence, the synchronous clock is a sort of *secondary clock*, where the *primary clock* is the power company's current.

Electronic clock: A sort of *electric clock* that does away with the mechanical system altogether. Instead, it uses an electric circuit itself as the timekeeping element. It is the capability to create all-electronic clocks that produced the atomic revolution in timekeeping.

Electronic oscillator: An electronic circuit that alternates between two states in a given time.

Electrostatic clock: A type of early *electric clock* that used dry piles, a type of early battery that worked without the need for any liquid component, to move a pendulum between two terminals.

Entrainment: The process by which two harmonic oscillators come to have the same period, such as in *antisynchronization*.

Epact: A method of *intercalation* for adjusting the lunar year to the solar year.

Ephemeris year: A time standard adopted by the International Astronomical Union in 1952 in which hours, minutes, and seconds are defined as fractions of the earth's journey around the sun.

Epicyclical gearing: Gears mounted so that the center of one travels around the center of another.

Epoch: The reference point from which time is counted.

Equal hours: The division of the day and night into twenty-four hours that are equal in length. Today, calculated as a fraction of the *mean solar year*.

Equation of time: From the Latin *aequatio*, "equalizing," a table of the difference between time as kept by the sun and time as kept by a clock.

Equinoxes: Days when light and darkness are equal in length in the spring and fall.

Escapement: A device to control the flow of energy into the timekeeping mechanism.

Escape wheel: The gear in an *escapement*.

Feria: Roman public holidays, which carried over into Catholicism as weekdays when believers had to attend mass.

Fusee: A cone-shaped pulley or axle with grooves in its surface to equalize the force of the spring driving a *spring-driven clock*. The fusee device is elegant in its simplicity: It is a cone-shaped pulley or axle with grooves in its surface. The drive train is wrapped around the fusee and as the spring unwinds it gives more or less mechanical advantage. Over time, by trial and error, clockmakers discovered the optimal spindle shape for a fusee.

Geng: A system of unequal hours in premodern China, which divided the day from dusk to dawn into five parts.

Global Positioning System: A system of satellites that, by broadcasting a time signal, allows users to triangulate their position in space.

Gnomon: The shadow-casting part of a sundial.

Grasshopper escapement: An *escapement* invented by John Harrison. The top arm and its pallets resist the escape wheel's turning, causing the hinge to open. At the same time, the second vertical arm and pallet pivot toward the wheel, catch, and push it back slightly to release the first arm and pallet, which then swing back to their start position. In effect, one arm "hands over" the escape wheel to the other in an action that looks like the "kicking" of a grasshopper's legs. Thanks to the weighting and balance of the arms, both the first pallet's reset and the second pallet's approach are automatic. Furthermore, there is little friction or wear and tear on the pallets, and the grasshopper escapement needs no lubrication.

Greenwich mean time: A single, universally valid time reference based on the local time in Greenwich, England.

Gregorian calendar: A correction to the Julian calendar made in 1582 CE.

Gridiron pendulum: A bimetallic pendulum invented by John Harrison to compensate for temperature fluctuations.

Haab': The Mayan term for a 260-day cycle in the Mesoamerican calendar.

Hairspring: See *balance spring*.

Harmonic oscillator: An object that, when disturbed from its equilibrium position, experiences a restoring force that is proportional to the displacement.

Hipp toggle: An *impulse clock* invented by Matthäus Hipp that uses an electromagnet to give a push to a clock pendulum when its swing drops below a certain level.

Horologium: In the Middle Ages, any observational device including a chart for determining the hour from astronomical observations or a sundial or water clock. It would eventually give its name to the word for "clock" in several European languages.

Horology: The science and technology of keeping time.

Hour: See *equal hours*, *unequal hours*, *canonical hours*.

Ides: Roughly the middle of the month in the Roman calendar. Usually the 13th or 15th—just so long as there were 16 days until the next *kalends*.

Impulse clock: An *electric clock* in which electricity powers the timekeeping mechanism. The electric clock of Matthäus Hipp (1815–1893), who was born in Germany but emigrated to Switzerland in 1852, is an example of an early impulse clock. Hipp's clock, which he invented around 1842, was also the first mass-market electric clock. It functioned by means of a Hipp toggle (also known as the butterfly escapement), which uses an electromagnet to give a push to a clock pendulum when its swing drops below a certain level; because the magnet does not physically touch the pendulum but only pushes it at the bottom of its arc when its kinetic energy is at its maximum, it hardly upsets its accuracy at all.*

Indiction: A 15-year cycle in the Roman *calendar* and dating system.

Inertial frame: The concept that an object's velocity depends on the measurer's own movement.

*On Hipp, see Helmut Kahlert, "Lorenz Bob und Matthäus Hipp," in *Alte Uhren und moderne Zeitmessung* (Munich: Callwey Verlag, 1987), 22:4; and Hans Rudolf Schmid, "Hipp, Matthäus," in *Neue Deutsche Biographie* (Berlin: Duncker & Humblot, 1972), 9:199–200.

Intercalation: The practice of adding a "leap" day, week, month, or other interval to reconcile *calendars* to natural cycles.

International Date Line: The imaginary line where the date shifts for travelers.

Islamic calendar: A strictly lunar *calendar* of 354 or 355 days.

Isochronism: The property of an event of always taking the same amount of time, independent of variables.

Italian hours: A system for reckoning the day in twenty-four hours that began at sunset. It lasted until the mid-eighteenth century.

Julian calendar: Correction to the Roman *calendar* introduced by Julius Caesar in 45 BCE and used until 1582 CE.

Kalends: The first of the month in the Roman *calendar*.

Ke: In premodern China, one of the divisions of the day from midnight to midnight.

Lever escapement: Invented by the English clockmaker Thomas Mudge in about 1755 with several improvements added through the nineteenth century. In a lever escapement, the *balance wheel* is attached to an axle, or roller. A pin on the roller pushes on the lever as the balance wheel causes the roller to rotate. The lever, in turn, rocks the pallets back and forth, which, as on the anchor escapement, interact with the teeth of the escape wheel and drive the clock's mechanism. The lever escapement has several advantages: it is very precise, it can be restarted easily, and it separates the balance wheel from the gearing, which reduces accuracy-reducing recoil.

Long Count Calendar: A period in the Mesoamerican calendar combining both base-20 and base-18 counting, dating *b'ak'tun* cycles (about 394 1/2 years) from the mystical creation date of August 11, 3114 BCE.

Lunar-distance method: A means of determining the time, and therefore longitude, from tables of the moon's observed distance from other celestial bodies.

Lunation: The cycle of the phases of the moon.

Mainspring: See *torsion spring*.

Master clock: See *primary clock*.

Mean solar day: An imaginary day of "average" length. Suggested by John Flamsteed (1646–1719).

Mean solar time: The time according to a clock that is calibrated to the *mean solar day*. Contrast with *apparent solar time*.

Merkhet: An ancient Egyptian device consisting of a hanging weight for sighting stars to tell time at night.

Mesoamerican calendar: An integral tradition that lasted from the Olmec and Zapotec cultures around 1800 BCE to European conquest in the sixteenth century CE, characterized by a base-20 system and the use of zero as a placekeeper.

Metonic year: The lowest common product of the solar year—that is, the amount of time it takes the sun to complete its journey through the seasons, from spring equinox to spring equinox or summer solstice to summer solstice—and the synodic month—that is, the period of the moon's phases. The Metonic year comes to about 19 solar years or 235 months.

Millennialism: Fear that the revolution of cosmic cycles of time will lead to a new order of the human world.

Nones: In the Roman calendar, the ninth day before the *kalends*.

Nundium: The Roman week of eight days.

Nuremberg egg: A type of early clock that could be worn.

Oblique ascension: The altitude of an imaginary point on the celestial equator.

Oscillator: A phenomenon that changes states at a known rate. See *electronic oscillator, harmonic oscillator*.

Piezoelectricity: The property of a solid material accumulating an electric charge when it is put under mechanical stress.

Pocket watch: A clock designed to be carried in a pocket.

PM: *Post meridiem,* "after noon"—*meridies* being the Latin for "noon."

Primary clock: An electric clock that controls one or more *secondary clocks*.

Primum mobile: In Ptolemaic astronomy, an invisible outermost sphere, the "first moved," which imparts motion to the rest of the spheres on which the planets move.

Process time: A computer science term that measures the work done by a computer's central processing unit (CPU).

Ptolemaic: A model of the cosmos that has the earth at (or near) the center of the universe.

Qi: A unit in Chinese calendrical timekeeping. A qi was defined as when the solar ecliptic longitude moved 15°. The solar year had 24 qi and 5.24 extra days.

Quartz clock: An *electronic clock* that uses a quartz crystal in its *electronic oscillator*.

Radio clock: A sort of *secondary clock* that functions off of a radio-transmitted time signal.

Relative time: The belief that time exists only in relation to observable moving objects.

Relativity: The scientific theory that space and time are not separate concepts (as Newton held), but intimately related—a continuum called "space-time."

Relativity of simultaneity: A corollary of the theory of *relativity* that states that, because observations are changed by the observer's *inertial frame*, events can't really happen at the same time.

Remontoire: A secondary drive that moves a clock mechanism and is in turn rewound at regular intervals by the main drive. Without a remontoire, mechanical resistance to the clock's mechanism will affect the drive and introduce inaccuracies.

Resolution: How small an increment of time a computer clock can count.

Rewound clock: An *electric clock* in which the timekeeping mechanism is mechanical but electricity is used to rewind the clock.

Right ascension: The "longitude" of any object on the *celestial sphere*.

Salah: In Islam, the five-times-daily prayer required for all believers.

Seasonal hours: See *unequal hours*.

Secondary clock: An electric clock synchronized to a *primary clock*.

Sempiternal: A thing that has a beginning but no end; for instance, while God is eternal, angels, which are immortal created beings, have a beginning but no end.

Sexagesimal: Base-60.

Sidereal day: The time it takes for the stars to seem to revolve in a complete circle and return to the same place. It lasts 23 hours, 56 minutes, and 4 seconds according to the modern reckoning of the hours as a fraction of the *mean solar day*.

Sidereal month: The time it takes for the moon to return to a given position against the stars.

Sidereal year: The time it takes the sun to travel around the ecliptic—that is, return to a given position against the stars.

Slave clock: See *secondary clock*.

Solar year: Another name for the tropical year.

Solenoid: A coil of wire that acts as an electromagnet when a current is sent through it.

Solstices: The shortest and longest days of the year. The longest day of the year is the summer solstice, while the shortest is the winter solstice.

Spring-driven clock: A clock wherein the power comes from an unwinding spring.

Stackfeed: A device to equalize the force of an unwinding spring in a *spring-driven clock*, mostly used in German clocks from around 1530 to 1640. It consists of a *cam*, or eccentrically shaped gear, that presses on the spring by means of a stiff metal arm. As the clock runs down, the cam turns, changing the amount of force opposing the spring. Thus, instead of changing the force coming from the spring, it regulates the amount of force the spring itself is exerting. This is obviously less efficient; however, a stackfeed was easier to make than a *fusee*.

Standard time: A synchronized time standard used throughout a region. See also *time zone*.

Strob escapement: Similar to a *virge-and-foliot* but, rather than a crown wheel, this relied on two wheels or gears fitted with precisely spaced pegs that alternately rotated a semicircular pallet attached to the virge and swung it back.

Synchronous clock: A type of electric clock invented by Henry Ellis Warren (1872–1957) in 1918 that relies on the power grid itself, using the frequency of the alternating current to keep time. (Alternating current reverses direction many times per second; direct current is a steady flow. Alternating current is a more efficient means of transmitting electricity, which is why it is used by utility companies.) Synchronous clocks have no oscillator; the electricity flowing through the electrical wires of a power grid is itself the timekeeping element. Because they rely on an external factor, these clocks are very accurate—unless there's a blackout.

Synodic month: The cycle of the phases of the moon.

Taylorism: Frederick Winslow Taylor's principles of efficient scientific time management in industrial production.

Time zone: A scheme for adjusting *standard time* for time differences as one moves east or west.

Torsion spring: A spring powering a *spring-drive clock* that derives its power from the untwisting of a piece of metal.

Tropical year: The time between equinox and equinox, or solstice and solstice.

Tropics: The northern and southern points of the ecliptic, where the sun seems to "turn" aside from its north–sound wandering on the solstices. The word is derived from the Greek for "turn." The northern point is called the Tropic of Cancer, the southern the Tropic of Capricorn. Though the tropics are named for the constellation in which the "turning" took place at the time they were named by the ancient Greeks, today they are in Taurus and Sagittarius.

Tuning-fork clock: An *impulse clock* invented by Alfred Niaudet-Bréguet that uses the frequency of an electrically driven tuning fork to control a small *escape wheel*.

Tzolk'in: Mayan term for a 365-day cycle in the Mesoamerican calendar.

Unequal hours: Divisions of the periods between sunrise and sunset into 12 parts each, which would be longer or shorter depending on the time of the year.

Virge-and-foliot escapement: A device that enabled the construction of the first all-mechanical clock. The "virge" part of the virge-and-foliot is named from the Latin *virga*, "stick" or "rod." The escapement itself, or "crown wheel," is a gear with vertical sawtooth-shaped teeth (thus the name "crown"). The virge has two tabs called "pallets" offset at such an angle that, as the crown wheel rotates thanks to the downward pull of the weights, the pallets will engage and rotate the virge, which in turn moves the "foliot," a weighted bar. A tooth on the opposite side then catches the other palette, rotating it back and returning the foliot to its original position. The virge-and-foliot serves to transform the downward pull of gravity into a regular oscillating motion, producing a characteristic "tick-tock" as it rotates forward and back. In other words, it transforms gravitational energy from the weights into a regular oscillating motion that enables the hands of the clock to thus move forward at a steady rate. Moreover, if the weights on the foliot are moved inward or outward, the period of the cycle can be adjusted, thus regulating the clock.

Wall clock time: In computer science, time as perceived by human beings.

Water clock: A timekeeping or timing device regulated by the flow of water into a basin.

Wristwatch: A timepiece that can be worn on the wrist.

Y2K problem: The fear of widespread computer malfunctions when the *epoch* changed to January 1, 2000.

Zenith: The highest point of a celestial object in its travel across the sky.

Zodiac: Mnemonic patterns for organizing the brightest stars in the sky, from the Greek *zōdiakos kyklos*, or "circle of animals."

One. Scholars and Spheres

1. Derek J. de Solla Price, "An Ancient Greek Computer," *Scientific American* 200, no. 6 (June 1959): 60–67; "Gears from the Greeks: The Antikythera Mechanism; A Calendar Computer from ca. 80 B.C.," *Transactions of the American Philosophical Society* 64, no. 7 (1974): 1–70.

2. Phillip Ball, "Complex Clock Combines Calendars," *Nature* 454, no. 7204 (July 30, 2008): 614–617. The best recent work on the Antikythera Mechanism is Alexander Jones, *A Portable Cosmos: Revealing the Antikythera Mechanism, Scientific Wonder of the Ancient World* (New York: Oxford University Press, 2017).

3. I would like to thank Professor Bert Hall of the University of Toronto for helping me formulate and clarify these ideas.

4. Jared Diamond, *Guns, Germs, and Steel: The Fates of Human Societies* (New York: W. W. Norton, 1999).

5. Ray Tomlinson's email history page, http://openmap.bbn.com/~tomlinso /ray/home.html, accessed September 8, 2008.

6. Augustine, *Confessions*, 11:14.

7. The question was perhaps first expressed by Karl Marx in an 1863 letter to Friedrich Engels, where he identified the clock as the first applied use of the machine, "from which the whole theory of production of regular motion was developed." *The Letters of Karl Marx*, ed. and trans. Saul K. Padover (Englewood Cliffs, NJ: Prentice-Hall, 1979), 168. It was famously advanced by Lewis Mumford: "The clock . . . is a piece of power machinery whose 'product' is seconds and minutes: by its essential nature it dissociated time from human events and helped created the belief in an independent world of mathematically measurable sequences: the special world of science." *Technics and Civilization* (New York: Harcourt, Brace, 1934), 15. Other critical (if dated) pieces on premodern ideas of time and which established the premodern world as a timekeeping "other" are Marc Bloch, *La sociètè fèodale*, 2 vols. (Paris: Albin, 1939–1940), 1:117; Lucien Febvre, "Temps flottant, temps dormant," in *Le problème de l'incroyance: La religion de Rabelais* (Paris: Albin, 1942), 426–434. Linking time measurement and medieval capitalism is Jacques Le Goff, "Merchant's Time and Church's Time," in *Time, Work, and Culture in the Middle Ages*, trans. Arthur Goldhammer (Chicago: University of Chicago Press, 1980), 29–42. For a counter to these, see Gerhard Dohrn-van Rossum's critique

in his extremely important *The History of the Hour*, trans. Thomas Dunlap (Chicago: University of Chicago Press, 1996). Finally, for a good historiographical overview of the social use of time, see Peter Burke, "Reflections on the Cultural History of Time," *Viator* 35 (2004): 617–626.

8. Gerald James Whitrow, *What Is Time?* (Oxford: Oxford University Press, 1972), *Time in History* (Oxford: Oxford University Press, 1989).

9. Whitrow, *Time in History*, 15.

10. Ken Mondschein and Denis Casey, "Time and Timekeeping," in *Handbook of Medieval Culture*, ed. Albrecht Classen (Berlin: De Gruyter, 2015) 3:1657–1679.

11. Michael Rappenglück, "The Anthropoid in the Sky: Does a 32,000 Years Old Ivory Plate Show the Constellation Orion Combined with a Pregnancy Calendar?," in *Symbols, Calendars and Orientations: Legacies of Astronomy in Culture; IXth Annual meeting of the European Society for Astronomy in Culture (SEAC)*, ed. Mary Blomberg, Peter E. Blomberg, and Göran Henriksson (Uppsala: Uppsala University, 2003), 51–55.

12. Incidentally, Avebury, also in Wiltshire, is an even bigger but less famous stone circle than Stonehenge.

13. See, for instance, A. César González-Garcia, "Carahunge: A Critical Assessment," in *Handbook of Archaeoastronomy and Ethnoastronomy*, ed. Clive L. N. Ruggles (New York: Springer Science+Business Media, 2004), 1453–1460; and Ivan Ghezzi and Clive Ruggles, "Chankillo: A 2300-Year-Old Solar Observatory in Coastal Peru," *Science* 315, no. 5816 (2007): 1239–1243.

14. Ulrich Boser, "Solar Circle," *Archaeology* 59, no. 4 (2006): 30–35.

15. James Evans, *The History and Practice of Ancient Astronomy* (Oxford: Oxford University Press, 1998), 16–17. The story of the survival of Babylonian observations is interesting: They have come down to us because, centuries later, after the Babylonians were conquered by the Assyrians, the great king Ashurbanipal (the biblical Asenappar) had them recopied into his library. In 612 BCE, about two decades after his death, Ashurbanipal's library was set on fire and collapsed when the Assyrian capital, Nineveh, was sacked during the rebellion that ended the empire. This turned out to be a good thing for us, since British archaeologists were able to dig up the library in the nineteenth century. Because ancient Mesopotamian "books" were written by making wedge-shaped cuneiform marks on clay tablets, which were then baked hard by the fire, Ashurbanipal's library survived the millennia surprisingly well, and experts were able to decipher much of it.

16. Miguel Ángel Molinero Polo, "A Bright Night Sky over Karakhamun: The Astronomical Ceiling of the Main Burial Chamber in TT 223," in *Tombs of the South Asasif Necropolis: Thebes, Karakhamun (TT 223), and Karabasken (TT 391) in the Twenty-Fifth Dynasty*, ed. Pischikova Elena (Cairo: American University in Cairo Press, 2014), 201–238.

17. Whitrow, *Time in History*, 28.

18. Evans discusses MUL.APIN in *History and Practice*, 5–11.

19. Evans, *History and Practice*, 8.

20. Hesiod, "Works and Days," in *Hesiod, the Homeric Hymns, and Homerica*, trans. Hugh G. Evelyn-White (Cambridge, MA: Loeb Classical Library, 1914), ll. 384, 564–571.

21. Evans, *History and Practice*, 23–25.

22. Evans, *History and Practice*, 23–25.

23. Evans, *History and Practice*, 49, 64–65.

24. Interestingly, though the tropics are named for the constellation in which the "turning" took place at the time they were named by the ancient Greeks, today, they are in Taurus and Sagittarius. Because the sidereal year (that is, the time it takes the sun to travel around the ecliptic) is about 11/26,000 times longer than the tropical year (about 20 minutes), the background stars seem to shift with respect to the ecliptic. In the third book of the *Almagest*, Ptolemy credits the Greek astronomer Hipparchus, who lived sometime in the second century BCE, with having discovered this phenomenon.

25. Most importantly for the development of Western timekeeping, the date of Easter, the holiest day in the Christian year, was calculated according to a Metonic calendar. I will discuss the Gregorian calendar and the calculation, or computus, of Easter, in more detail in chapter 3.

26. *Naturalis Historia*, 7.213.

27. Evans, *History and Practice*, 131–132.

28. Dohrn-van Rossum, *The History of the Hour*, 29ff.

29. John Cassian, *De coenobiorum institutis*, 2:17, in *Patrologia Latina*, ed. Jacques Paul Migne (Paris, 1846), 49:108.

30. See Daniel F. Mansfield and N. J. Wildberger, "Plimpton 322 Is Babylonian Exact Sexagesimal Trigonometry," *Historia Mathematica* 44, no. 4 (November 2017): 395–419. Uncorrected proof available at https://doi.org/10 .1016/j.hm.2017.08.001, accessed August 24, 2017.

31. *Asser's Life of King Alfred*, trans. Lionel Cecil Jane (London: Chatto & Windus, 1908), 86.

32. Abelard, letter 8, *The Letters of Abelard and Heloise*, trans. Betty Radice (New York: Penguin Classics, 2003).

33. Paris, Bibliothèque de l'Arsenal, 1186.

34. Guillaume Durand, *Rationale Divinorum Officorum* (Naples: Apud Josephum Dura Bibliopolam, 1859), iv.

Two. Cities and Clocks

1. Church bells (Latin *campana* or *clocca* from the Irish *clog*, whence French *cloche*, whence the English *clock*), are of unquestionably ancient date. Guillaume Durand, in book 4 of his *Rationale Divinorum Officorum* (Naples: Apud Josephum Dura Bibliopolam, 1859), attributes the custom of ringing bells to Pope Sabinian (604–606).

2. For leatherworkers, see René de Lespinasse and François Bonnardot, *Les métiers et corporations de la Ville de Paris: Le livre des métiers d'Etienne Boileau* (Paris: Imprimerie Nationale, 1879), 183; for mailmakers, see Durand, *Rationale divinorum officorum*, 56.

3. Gerhard Dohrn-van Rossum, *The History of the Hour*, trans. Thomas Dunlap (Chicago: University of Chicago Press, 1996), 218–219.

4. Heinrich Denifle, *Chartularium Universitatis Parisiensis*, 7 vols., ed. Heinrich Denifle, Emile Chatelain, Charles Samaran, and Emile A. van Moé (Paris: Delalain, 1889–1897), 1:546.

5. *Ordonnances des roys de France de la troisième race* (Paris: Imprimerie Royale, 1723–1849), 2:79.

6. Also see Lynn Thorndike, "Invention of the Mechanical Clock about 1271 AD," *Speculum* 16 (1941): 242–243.

7. Dohrn-van Rossum, *The History of the Hour*, 54. See also G. G. Meersseman and E. Adda, *Manuale di computo con ritmo mnemotecnico dell'arcidiacono Pacifico di Verona* (Padua: Herder, 1966). Pacificus's epitaph in the cathedral of Verona reads:

Horologium nocturnum nullus ante viderate
En invenit argumentum et priumus fundaverat
Horologioque carmen sphera coeli optimum,
Plura alia graviaque prudens invenit.

Johann Beckmann and William Johnston, *A History of Inventions and Discoveries*, trans. William Johnston (London: Longman, 1817), 1:424.

8. Dohrn-van Rossum, *The History of the Hour*, 59.

9. See Charles Homer Haskins, "Nimrod the Astrologer," in *Studies in Medieval Science* (Cambridge, MA: Harvard University Press, 1924), 336–345; and David Juste, "On the Date of the Liber Nemroth," *Journal of the Warburg and Courtauld Institutes* 67 (2004): 255–257.

10. Nimrod, *Liber de astronomica*, Paris, BnF, lat. 14754, fols. 203r–238v.

11. Nimrod, *Liber de astronomica*, Paris, BnF, lat. 14754, fols. 203r–238v.

12. A. R. Green, *Sundials, Incised Dials or Mass-Clocks* (London: Society for Promoting Christian Knowledge, 1926).

13. Allan A. Mills, "Seasonal-Hour Sundials on Vertical and Horizontal Planes, with an Explanation of the Scratch Dial," *Annals of Science* 50, no. 1 (January 1993): 83.

14. Green, *Sundials*.

15. David King, *In Synchrony with the Heavens* (Leiden: Brill, 2004), 1:855–74. The essential work on Arabic astronomy remains George Saliba's *A History of Arabic Astronomy: Planetary Theories During the Golden Age of Islam* (New York: New York University Press, 1994). European astronomical instruments and theory have been studied in detail by Emmanuel Poulle in his several works, especially his *Les instruments astronomiques du Moyen Age* (Paris: Société internationale de l'astrolabe, 1983), *Les instruments de la théorie des planètes selon Ptolémée: Équatoires et horlogerie planétaire du XIIIᵉ au XVIᵉ siècle*, 2 vols. (Geneva: Droz, 1980–1981), *Les sources astronomiques (textes, tables, instruments)* (Turnhout, Belgium: Brepols, 1984), *Les tables alphonsines, avec les canons de Jean de Saxe* (Paris: Editions du Centre national de la recherche scientifique, 1987–1988), and his more recent

translation of Giovanni Dondi dell'Orologio's *Tractatus astrarii* (Geneva: Droz, 2003).

16. J. Millàs Vallicrosa, *Assaig d'història de les idees físiques i matemàtiques a la Catalunya medieval* (Barcelona: Edicions Científiques Catalanes, 1931); Thomas Glick, "Éloge: José María Millás Vallicrosa (1897–1970) and the Founding of the History of Science in Spain," *Isis* 68, no. 2 (June 1977): 276–283.

17. Gerbert of Aurillac, *De utilitatibus astrolabii*, ed. N. Bubnov, in *Gerberti opera mathematica* (Berlin: Berolini, 1899), 109–147. See also Marco Zuccato, "Gerbert of Aurillac," in *Medieval Science, Technology, and Medicine*, ed. Thomas Glick, Steven John Livesey, and Faith Wallis (New York: Psychology Press, 2005), 192–194. For medieval knowledge of climate bands, see below on the textural tradition of the *Somnium Scipionis*.

18. Dohrn-van Rossum, *The History of the Hour*, 79; for Hermann's work, see *Annales Bertoldi*, ed. G. H. Pertz, in *Monumenta Germaniae historica, scriptores* (Turnhout, Belgium: Brepols, 1844), 5:268.

19. Guillaume de Saint-Pathus, *Recueil des historiens de Gaul et de France* XX, 71–73.

20. Christine de Pisan, *Le Livre des Faits et Bonnes Meurs du Sage Roy Charles V*, vol. 1, ed. Suzanne Solente (Paris: H. Champion, 1936), 56.

21. Dino Compagni, *Crónica*, vol. 2 p. 19, ed. Isidoro del Lungo (Paris, 1891), 98–99.

22. Dohrn-van Rossum, *The History of the Hour*, 5.

23. The image is from an illuminated Hebrew Bible known as the Coburg Pentateuch, London, British Library, Add. 19776, fol. 72v.

24. See Carol Herselle Krinsky, "Seventy-Eight Vitruvius Manuscripts," *Journal of the Warburg and Courtauld Institutes* 30 (1967): 36–70.

25. For instance, in the fifth century, Macrobius mentions using a water clock for astronomical observations. In the year 507, Theodoric the Great, king of the Ostrogoths, the Germanic people who had conquered Italy, sent a sundial and water clock to King Gundobald of the Burgundians. Theodoric's minister, the scholar Cassiodorus, praised these devices as necessary for monastic life. Bede, in his *On the Reckoning of Time* (*De temporum ratione*), discusses the use of a *horologium*, which is presumably a water clock, to help find the vernal equinox. *De temporum ratione* 30, in *Patrologia Latina*, ed. Jacques Paul Migne (Paris, 1862), 90:430. Likewise, in a ninth-century commentary on the Benedictine Rule attributed to Abbot Hildemar of Corbie in modern France, we find water clocks mentioned as a necessity for determining the divisions of the night in which to pray. Hildemar says specifically that one ought to mark hours that varied according to the length of day or night (that is, seasonal hours) and that, to keep track of these, one will need a water clock. *Vita et regula SS. P. Benedicti una cum expositione regulae a Hildemaro tradita*, ed. Rupert Mittermüller (Regensberg, 1880), 8:277–278.

26. Finbarr Barry Flood, *The Great Mosque of Damascus: Studies on the Makings of an Umayyad Visual Culture* (Boston: Brill, 2000), 116, 121.

27. See Ismail al-Jazari, *The Book of Ingenious Devices*, trans. Donald Hill (Boston: Kluwer Academic, 1979); and Donald Hill, *Arabic Water Clocks* (Aleppo: University of Aleppo, 1981).

28. *Chronica regia Coloniensis continuatio IV*, ed. G. Waitz, in *Monumenta Germaniae historica, Script. Rer. Germ.* 18 (1880), 263; and *Conradi de Fabaria Casus Galli*, ed. G. H. Pertz, *Monumenta Germaniae historica, scriptores* 2 (1829), 178. Dohrn-van Rossum, *The History of the Hour*, gives a good summary on 73–74.

29. *Acta sanctorum* (Antwerp, July 2, 1721), 155.

30. Guillaume of Auvergne, *De anima* I, 7a.

31. *Acta sanctorum* (Paris, October 13, 1883).

32. *Acta sanctorum* (Antwerp, May 4, 1685), 401.

33. Lynn White Jr., *The Sphere of Sacrobosco and Its Commentators* (Chicago, 1949), 229–230. The translation is White's.

34. Dohrn-van Rossum, *The History of the Hour*, 81–83.

35. For a full discussion of the clock's workings and feasibility, as well as a translation, see A. A. Mills, "The Mercury Clock of the *Libros del Saber*," *Annals of Science* 45, no. 4 (July 1988): 329–344.

36. For a full accounting, see Dohrn-van Rossum, *The History of the Hour*, 91–94.

37. John D. North wrote several works on Wallingford, of which the best and most comprehensive is *God's Clockmaker* (New York: Bloomsbury Academic, 2005); see also North's translations of Wallingford's treatises, *Richard of Wallingford* (Oxford: Clarendon, 1976).

38. *Recueil des historiens des Gaules et de la France* XXIII, 410.

39. Steven A. Epstein, "Business Cycles and the Sense of Time in Medieval Genoa," *Business History Review* 62 (1988): 238–260.

40. Dohrn-van Rossum, *The History of the Hour*, 129–153.

41. Guillaume Durand, trans. Jean Golein, BNF MS fr 176 (variant readings from BNF MS fr 931):

> Que pape Savinien ordena que on sonast les cloches aus heures du iour
> par les eglises aus xii heures. Et ce a ordené le roy charles primier á Paris
> le cloches qui á chascune heure comment per poins á manere d'arloges
> [orologes] si comme il apiert en a son palais et au boys et á saint Pol.
> Et a fait venir ouveriers destrange pays á gravis frès pour ce faire, afin que
> religieus et autres gens sachent le heures at aient propres manières et
> devocion di ipur et de nuit [pour Dieu servier]. Comment que, par devant,
> on sonnast une fois á prime et deux [trois] fois á tierce, si n'avoit-on mie
> si certaine congnoissance des heures comme on a, et peut-on dire d'icelui
> Charles le VIᵉ [Vᵉ] roy de France que *sapiens dominabitur astris*; car luise
> le souliel ou non, l'on scet tousjours les heures sans defailli, par icelles
> cloches atrampes [atrempées]. Et devons savoir qu'il a en l'eglise cinq
> maniéres de cloches; c'est assavoir esquelles, timbres, noles et noletes et

cloches. La cloche sonne en l'eglise, l'esquelle en refectourer, le timbre ou cloistre, la nole ou choeur, la nolete en l'orologe.

I have reproduced some of the diacriticals added in nineteenth-century sources to aid comprehension.

42. Dohrn-van Rossum, *The History of the Hour*, 218–220. Febvre and Le Goff both looked at this phenomenon to establish their arguments. Febvre, however, reinforces his idea of medieval "lived time" by concentrating on the lack of precision and the human element in the system of looking at the clock and then physically hitting a bell with a hammer, and Le Goff argues for his "merchant's time" by citing the urge to measure by an authoritative machine in the first place.

For two excellent works on changing time regimes in England, see Chris Humphrey, "Time and Urban Culture in Late Medieval England," in *Time in the Medieval World*, ed. Chris Humphrey and W. Mark Ormrod (Woodbridge: York Medieval Press, 2001), 105–118; and Karen Smyth, "Changing Times in the Cultural Discourse of Late Medieval England," *Viator* 35 (2004): 435–454.

43. *Mémoires de la Société de l'Histoire de Paris et de l'Ile-de-France* 17 (1892): 60–64. (*Not* 1891 as Dohrn-van Rossum has it in his notes!)

44. *Ordonnances* 1:728.

45. Lespinasse and Bonnardot, *Métiers*, 3:106.

46. Du Boulay, 3:451.

47. William of Ockham, *Phil. Nat.*, 4:7. William of Ockham's *Opera philosophica et theologica*, edited by Gedeon Gál et al., is available in 17 volumes from the Franciscan Institute (St. Bonaventure, NY, 1967–1988); his *Expositio in libros Physicorum Aristotelis* books 4–8 is volume 5 (1985). His *summa* of the *Physics* has been translated by Julian Davies as *Ockham on Aristotle's Physics: A Translation of Ockham's Brevis Summa Libri Physicorum* (St. Bonaventure, NY: Franciscan Institute, 1989). The preeminent discussion of Ockham on time remains Herman Shapiro's *Ockham on Motion, Time and Place* (St. Bonaventure, NY: Franciscan Institute, 1957), on which I lean heavily in this discussion. Shapiro mainly argues from the *Summula philosophia naturalis*; I will incorporate details from his other works as needed. Also of use are André Goddu's *The Physics of William of Ockham* (Leiden: Brill, 1984) and Calvin G. Normore's 1975 University of Toronto PhD dissertation, "The Logic of Time and Modality in the Later Middle Ages: The Contribution of William of Ockham."

48. Jean Buridan, *Quaestiones super octo Physicorum libros Aristotelis* (Paris, 1509); reprinted as *Kommentar zur Aristotelischen Physik* (Frankfurt: Minerva, 1964), 4:12, 78v–79r.

49. The original Latin poem and a translation may be found in Nicole Oresme, *Le livre du ciel et du monde*, ed. Albert D. Menut and Alexander J. Denomy (Madison: University of Wisconsin Press, 1968), 580–581.

50. Nicholas of Cusa, *De visione*, trans. Jasper Hopkins (Minneapolis: Arthur J. Banning, 1988), 700.

51. Dohrn-van Rossum, *The History of the Hour*, 228.

52. Dohrn-van Rossum, *The History of the Hour*, 233.

53. Dohrn-van Rossum, *The History of the Hour*, 220–221.

54. Dohrn-van Rossum, *The History of the Hour*, 232.

55. *Acta sanctorum* (Antwerp, March 1, 1668), 575: "Nam Sacristaria conuentus, quæ pulsare pro Matutino debebat circa noctis medium, excitata fuit inter nonam & decimam horam, & existimans quod esset media nox, pulsauit de facto pro dicto Matutino."

Three. Savants and Springs

1. Galileo Galilei, *Dialogues concerning Two New Sciences*, trans. Henry Crew and Alfonso de Salvio (Norwich, NY: William Andrew, 1914), 178.

2. In his appendix to *De beghinselen der weeghconst* (*The Principles of the Art of Weighing*) (Leiden, 1586).

3. Nicholas of Cusa, *Opera omnia* (Leipzig: Meiner, 1937), 5:221–222, 233ff.

4. Stillman Drake's work, such as his *Discoveries and Opinions of Galileo* (Garden City, NY: Doubleday, 1957) and *Galileo at Work: His Scientific Biography* (Chicago: University of Chicago Press, 1978) is still fundamental in this field.

5. See Domenico Bertoloni Meli, *Thinking with Objects: The Transformation of Mechanics in the Seventeenth Century* (Baltimore: Johns Hopkins University Press, 2006), 131–134; Christopher M. Graney, *Setting Aside All Authority: Giovanni Battista Riccioli and the Science against Copernicus in the Age of Galileo* (South Bend, IN: University of Notre Dame Press, 2015).

6. Arturo Castiglione, "The Life and Work of Santorio (1561–1636)," trans. Emilie Recht, *Medical Life* 38 (1931): 729–785.

7. Probably the best work on his life is physicist C. D. Andriesse's *Huygens: The Man behind the Principle*, trans. Sally Miedema (Cambridge: Cambridge University Press, 2005).

8. See Hans van den Ende, *Huygens' Legacy: The Golden Age of the Pendulum Clock* (Castletown: Fromanteel, 2004).

9. See Michael Headrick, "Origin and Evolution of the Anchor Clock Escapement," *Control Systems Magazine* 22, no. 2 (2002): 41–52.

10. See Lynn White Jr., *Medieval Technology and Social Change* (New York: Oxford University Press, 1966), 126–127; and Dohrn-van Rossum, *The History of the Hour*, 121.

11. See the entry on Jost Bürgi, in Lance Day and Ian McNeil, eds., *Biographical Dictionary of the History of Technology* (New York: Routledge, 1996), 204–205.

12. See Dohrn-van Rossum, *The History of the Hour*, 122.

13. Dohrn-van Rossum, *The History of the Hour*, 123.

14. Samuel Pepys, *Diary and correspondence of Samuel Pepys: The diary deciphered by Rev. J. Smith from the original shorthand ms. Life and notes by Richard, Lord Braybrooke* (New York: Bigelow, 1905), 235.

15. On the lost minutes, see Robyn Adams and Lisa Jardine, "The Return of the Hooke Folio," *Notes and Records of the Royal Society of London* 60, no. 3 (September 22, 2006): 235–239.

16. Rictor Norton, "The Gregorian Underworld," http://rictornorton.co.uk /gu11.htm, accessed August 8, 2017.

17. Jacob Rosenbloom, "The History of Pulse Timing with Some Remarks on Sir John Floyer and His Physician's Pulse Watch," in *Annals of Medical History* (New York: Paul B. Hoebler, 1922), 4:98.

18. The comprehensive work on this is Elly Truitt's *Medieval Robots: Mechanism, Magic, Nature, and Art* (Philadelphia: University of Pennsylvania Press, 2015).

19. Elly Truitt, "Temporal Media: The Codex and the Clock in Late Antiquity and the Middle Ages" (invited lecture given at the Five College Medieval Studies Seminar, Smith College, November 15, 2017).

20. David S. Landes, *Revolution in Time* (Cambridge, MA: Harvard University Press, 1983), 121; I have rechecked the translation.

21. For antecedents, see Milič Čapek, "The Conflict between the Absolutist and the Relational Theory of Time before Newton," *Journal of the History of Ideas* 48 (1987): 595–608.

22. Milič Čapek, "The Conflict between the Absolutist and the Relational Theory of Time before Newton," *Journal of the History of Ideas* 48 (1987): 595–608.

23. Isaac Newton, *Philosophiae Naturalis Principia Mathematica,* trans. Andrew Motte (1729), rev. Florian Cajori (Berkeley: University of California Press, 1934), 6–12.

24. For a background of the political milieu, and how Newtonianism came to be associated with the party of the *philosophes*, see J. B. Shank's *The Newton Wars* (Chicago: University of Chicago Press, 2008).

Four. Navigators and Regulators

1. The best-known modern treatment of the longitude problem, if not a scholarly one, is, of course, Dava Sobel's *Longitude: The True Story of a Lone Genius Who Solved the Greatest Scientific Problem of His Time* (New York: Walker, 1995). David S. Landes, in his *Revolution in Time* (Cambridge, MA: Harvard University Press, 1983), also discusses Harrison. My task here is, of course, to retell this well-known story in an accessible and entertaining manner for those who might be hearing it for the first time.

2. Sobel, *Longitude*.

3. Wolfgang Köberer, "On the First Use of the Term 'Chronometer,'" *Mariner's Mirror* 102, no. 2 (2016), 203–206.

4. On the much- (and unjustly) vilified Maskelyne, see Derek Howse, *Nevil Maskelyne: The Seaman's Astronomer* (Cambridge: Cambridge University Press,1989); and also his *Greenwich Time and the Longitude*, rev. ed. (London: Philip Wilson, 2003).

5. J. J. O'Connor and E. F. Robertson, "Longitude and the Académie Royale," MacTutor History of Mathematics archive, http://www-groups.dcs.st -and.ac.uk/~history/HistTopics/Longitude1.html.

6. And one that has inspired writers to this day—see, for instance, Umberto Eco's *The Island of the Day Before*.

7. Described in the auction catalog of Dr. Crott Auctioneers, http://www .uhren-muser.com/cn/documents/News_7-2013_US.pdf, accessed August 24, 2017.

8. Jean Richer is little attested, but J. J. O'Connor and E. F. Robertson of the University of St. Andrews have his biography at http://www-history.mcs.st -andrews.ac.uk/Biographies/Richer.html, accessed August 24, 2017.

9. See Frederick W. Sawyer III, "Of Analemmas, Mean Time and the Analemmatic Sundial," *Bulletin of the British Sundial Society* 94, no. 2 (June 1994): 2–6; and 95, no. 1 (February 1995): 39–44.

10. For a biography, see David H. Clark and Stephen H. P. Clark, *Newton's Tyranny: The Suppressed Scientific Discoveries of Stephen Gray and John Flamsteed* (New York: Freeman, 2001).

11. The most recent biography of Halley is *Alan H. Cook, Edmond Halley: Charting the Heavens and the Seas* (Oxford: Clarendon Press, 1998). See also David W. Hughes, "Edmond Halley, Scientist," *Journal of the British Astronomical Association* 95, no. 5 (August 1985): 193–204, on his relationship with Flamsteed.

12. Quoted in Sobel, *Longitude*, 118.

13. In 1841, the American inventor Aaron Crane came up with a clock that used the torsion pendulum invented by Robert Leslie in 1793. The German Anton Harder reinvented the torsion clock in around 1880. The idea was that these clocks could run for a year, but their accuracy was notoriously bad.

14. Probably the most important theoretical work on the material in this section is E. P. Thompson's "Time, Work-Discipline, and Industrial Capitalism," *Past and Present* 38 (1967): 56–97.

15. Max Weber, *The Protestant Ethic and the Spirit of Capitalism*, trans. Talcott Parsons (New York: Routledge, 1992), 104.

16. And I do mean men: Only men had pockets!

17. Landes, *Revolution in Time*, 6–7.

18. William Radcliffe, *Origin of the New System of Manufacture, Commonly Called Power Loom Weaving* (London, 1828), 9–10, 59–67; reprinted in J. F. C. Harrison, *Society and Politics in England, 1780–1960* (New York: Harper & Row, 1965), 58–61.

19. *The Letters of Karl Marx*, ed. and trans. Saul K. Padover (Englewood Cliffs, NJ: Prentice-Hall, 1979), 168.

20. Thompson, "Time, Work-Discipline, and Industrial Capitalism," 56–97.

21. Steven A. Epstein, "Business Cycles and the Sense of Time in Medieval Genoa," *Business History Review* 62 (1988): 238–260.

22. Landes, *Revolution in Time*, 3.

23. Many books have been written on Lowell, but see especially Thomas Dublin, *Women at Work: The Transformation of Work and Community in Lowell, Massachusetts 1826–1860* (New York: Columbia University Press, 1979).

24. Melvin Dubofsky and Joseph A. McCartin, eds., *American Labor: A Documentary History* (New York: Palgrave Macmillan 2004), 59.

25. For a basic biography, see Frank Barkley Copley, *Frederick W. Taylor, Father of Scientific Management* (New York: Harper and Brothers, 1923).

26. Frederick Taylor, *Principles of Scientific Management* (New York: Harper & Brothers, 1913), 7.

27. There are a number of works on the Gilbreths and especially Lillian Gilbreth as a female academic pioneer, but see especially her autobiography, *As I Remember: An Autobiography* (Norcross, GA: Management and Engineering Press, 1998); and Frank Bunker Gilbreth Jr. and Ernestine Gilbreth Carey, *Cheaper by the Dozen* (New York: Thomas Y. Crowell, 1948, later made into a film), and *Belles on Their Toes* (New York: Thomas Y. Crowell, 1950).

28. For a good and readable brief history of Liebig's beef extract, see Clay Cansler, "Where's the Beef?," *Chemical Heritage Foundation*, Fall 2013/Winter 2014, https://www.chemheritage.org/distillations/article/where%E2%80%99s -beef, accessed August 24, 2017.

29. Vanessa Ogle, *The Global Transformation of Time: 1870–1950* (Cambridge, MA: Harvard University Press, 2015).

30. Alvin Powell, "America's First Time Zone," *Harvard Gazette*, November 10, 2011, http://news.harvard.edu/gazette/story/2011/11/americas -first-time-zone/, accessed August 24, 2017.

31. The history of electric clocks has mostly been written by amateur enthusiasts. Michel Viredaz's clock history site at http://www.chronometrophilia .ch/Electric-clocks/english.htm, accessed August 24, 2017, was very useful in preparing this section, as was the ClockDoc website at http://wp.clockdoc.org/, accessed August 24, 2017. On Zamboni, see Massimo Tinazzi, "Perpetual Locomotive of Giuseppe Zamboni," http://www.brera.unimi.it/sisfa/atti/1996 /tinazzi.html, accessed August 24, 2017; this page is quite scholarly and has an extensive bibliography.

32. Brian Bowers, *Sir Charles Wheatstone FRS: 1802–1875* (London: Science Museum, 2001), 154–155.

33. On Jones's system, see David Glasgow, *Watch and Clock Making* (London: Cassell, 1885), 314–315. See also Charles Piazzi Smyth, *Astronomical Observations Made at the Royal Observatory, Edinburgh* (Edinburgh: Neill, 1871), 13:xxxi–xxxii; F. Hope Jones, *Electrical Timekeeping* (London: NAG Press, 1940), https://clockdoc.org/gs/handler/getmedia.ashx?moid=23814&dt =3&g=1, accessed August 24, 2017; J. Kieve, *The Electric Telegraph: A Social and Economic History* (Newton Abbot, Devon: David & Charles, 1973). Bain himself wrote *A Short History of the Electric Clocks* (London: Chapman and Hall, 1852; repr. London: Turner & Devereux, 1973). A great website with animations that helped me immensely was J. E. Bosschieter's "History of the

Evolution of Electric Clocks," http://www.electric-clocks.nl/clocks/en/index.htm, accessed August 24, 2017.

34. On Dowd, see R. Newton Mayall, "The Inventor of Standard Time," *Popular Astronomy* 50 (April 1942): 204–208.

35. William F. Allen's story had been told many times; his papers are housed at the New York Public Library; see http://archives.nypl.org/mss/53, accessed August 24, 2017. The Smithsonian also has a good page at http://americanhistory .si.edu/ontime/synchronizing/zones.html, accessed August 24, 2017.

36. See Sandford Fleming's *Universal or Cosmic Time by Sandford Fleming, C.E., C.M.G., etc : Together with Other Papers, Communications and Reports in the Possession of the Canadian Institute respecting the Movement for Reforming the Time-System of the World and Establishing a Prime Meridian as a Zero Common to All Nations* (Toronto: Council of the Canadian Institute, 1885).

37. Ogle, *Global Transformation*, chapter 1 passim.

38. On cows, see Ogle, *Global Transformation*, 60.

39. Ogle, *Global* Transformation, 76–78.

40. Ogle, *Global Transformation*, 89.

41. Ogle, *Global Transformation*, 93.

42. Ogle, *Global Transformation*, 123–124.

43. On this, see Yulia Frumer's excellent *Making Time* (Chicago: University of Chicago Press, 2018).

Five. Rationalization and Relativity

1. I am not the first historian to make this comparison. See, for instance Stephen Kern's comparisons between Einstein and Durkheim in his *The Culture of Time and Space, 1880–1918*, 2nd ed. (1983; Cambridge, MA: Harvard University Press, 2003), 19–20.

2. There are a number of accessible explanations of Einstein for general readers, but best is Einstein's own *Relativity: The Special and the General Theory* (Princeton, NJ: Princeton University Press, 2015). Robert Geroch's *General Relativity from A to B* (Chicago: University of Chicago Press, 1981) is also good, while my personal favorite is the illustrated *Introducing Relativity,* written by Bruce Bassett and illustrated by Ralph Edney (Flint, MI: Totem Books, 2002).

3. Poul Anderson, *Tau Zero* (New York: Doubleday, 1970).

4. On Beeckman, see Klaas van Berkel, *Isaac Beeckman on Matter and Motion: Mechanical Philosophy in the Making* (Baltimore: Johns Hopkins University Press, 2013).

5. On Rømer, see Laurence Bobis and James Lequeux, "Cassini, Rømer and the velocity of light," *Journal of Astronomical History and Heritage* 11, no. 2 (2008): 97–105. The seminal paper is I. Bernard Cohen, "Roemer and the First Determination of the Velocity of Light (1676)," *Isis* 31, no. 2 (1940): 327–379.

6. Mari E. W. Williams, "Bradley, James (bap. 1692, d. 1762)," in *Oxford Dictionary of National Biography*, ed. H. C. G. Matthew and Brian Harrison (Oxford: Oxford University Press, 2004).

7. Hippolyte Fizeau, "Sur les hypothèses relatives à l'éther lumineux," *Comptes rendus hebdomadaires des séances de l'Académie des sciences* 33 (1851): 349–355.

8. Dorothy Michael Livingston, *The Master of Light: A Biography of Albert A. Michelson* (Chicago: University of Chicago Press, 1973).

9. Keith Davy Froome and Louis Essen, *The Velocity of Light and Radio Waves* (New York: Academic Press, 1969). For biographical information on Essen, see Alan Cook, "Louis Essen, O. B. E. 6 September 1908–24 August 1997," *Biographical Memoirs of Fellows of the Royal Society* 44 (1997): 143.

10. Ken Evenson et al., "Speed of Light from Direct Frequency and Wavelength Measurements of the Methane-Stabilized Laser," *Physical Review Letters* 29, no. 19 (November 6, 1972): 1346.

11. Arthur Eddington, *Report on the Relativity Theory of Gravitation* (London: Fleetway Press, 1918).

12. The claims that Eddington selected his data were made by John Earman and Clark Glymour, "The Gravitational Red Shift as a Test of General Relativity: History and Analysis," *Studies in History and Philosophy of Science* 11 (1980): 175–214; Daniel Kennflick has since sought to vindicate him in "Not Only Because of Theory: Dyson, Eddington and the Competing Myths of the 1919 Eclipse Expedition," in *Einstein and the Changing Worldviews of Physics*, ed. Christoph Lehner, Jürgen Renn, and Matthias Schemmel, vol. 12 of *Einstein Studies* (Boston: Birkhäuser, 2012).

13. Allie Vincent Douglas, *The Life of Arthur Stanley Eddington* (London: T. Nelson, 1956), 44.

14. http://www.gps.gov/systems/gps/, accessed July 20, 2017.

15. Einstein's legacy and relationship to modernity has been dissected many times. For varied perspectives and bibliography on this, see Peter L. Galison, Gerald Holton, and Silvan S. Schweber, eds., *Einstein for the 21st Century: His Legacy in Science, Art, and Modern Culture* (Princeton, NJ: Princeton University Press, 2008), and Gerald Holton and Yehuda Elkana, eds., *Albert Einstein, Historical and Cultural Perspectives: The Centennial Symposium in Jerusalem* (Princeton, NJ: Princeton University Press, 1984).

16. Bertrand Russell, *The ABCs of Relativity* (New York: Allen & Unwin, 1958), 134.

17. Russell, *ABCs of Relativity*, 136.

18. *Henry Ellis Warren: A Biographical Memoir* (New York: American Historical, n.d.), http://www.telechron.com/telechron/warren_bio.pdf, accessed July 20, 2017.

19. Alfred Niaudet-Bréguet, "Application du diapapason à l'horlogerie," in *Comptes rendus de l'Académie des Sciences* 63 (Paris, 1866): 991–992; see also Françoise Birck and André Grelon, eds., *Un siècle de formation des ingénieurs électriciens: Ancrage local et dynamique européenne; L'exemple de Nancy* (Paris: Éditions de la Maison des sciences de l'homme, 2006), 407.

20. Warren P. Mason, "Professor Walter G. Cady's Contributions to Piezoelectricity and What Followed from Them," *Journal of the Acoustical*

Society of America 58, no. 2 (August 1975): 301–309; Shaul Katzir, "From Ultrasonic to Frequency Standards: Walter Cady's Discovery of the Sharp Resonance of Crystals," *Archive for History of Exact Sciences* 62, no. 5 (September 2008): 469–487.

21. Warren Marrison and J. W. Horton, "Precision Determination of Frequency," *Proceedings of the Institute of Radio Engineers* 16, no. 2 (February 1928): 137–154; Warren Marrison, "The Evolution of the Quartz Crystal Clock," *Bell System Technical Journal* 27, no. 3 (July 1948), 510–588.

22. On the development of wristwatches, see Carlene Stephens and Maggie Dennis, "Engineering Time: Inventing the Electronic Wristwatch," *British Journal for the History of Science* 33, no. 4 (2000): 477–497. On the development of watches in general, see the epilogue to Alexis McCrossen, *Marking Modern Times: A History of Clocks, Watches, and Other Timekeepers in American Life* (Chicago: University of Chicago Press, 2016).

23. Willem de Sitter, "On the Secular Accelerations and the Fluctuations of the Longitudes of the Moon, the Sun, Mercury and Venus," *Bulletin of the Astronomical Institutes of the Netherlands* 4 (1927): 21–38.

24. Andrea Sella, "Essen's Clock," *Chemistry World*, January 27, 2014, https://www.chemistryworld.com/opinion/essens-clock/7017.article, accessed July 21, 2017. The paper was Louis Essen and J. V. L. Parry, "An Atomic Standard of Frequency and Time Interval: A Cæsium Resonator," *Nature* 176, no. 4476 (1955): 280.

25. Glenn K. Nelson, Michael A. Lombardi, and Dean T. Okayama, "NIST Time and Frequency Radio Stations: WWV, WWVH, and WWVB" (2005), http://tf.nist.gov/timefreq/general/pdf/1969.pdf, accessed July 21, 2017.

26. Ulrich Windl, David Dalton, Marc Martinec, and Dale R. Worley, "The NTP FAQ and HOWTO: Understanding and Using the Network Time Protocol (A First Try on a Non-technical Mini-HOWTO and FAQ on NTP)," https://www.eecis.udel.edu/~ntp/ntpfaq/NTP-a-faq.htm, accessed July 21, 2017.

27. Robert W. Bemer, "Time and the Computer," *Interface Age Magazine* 4, no. 2 (February 1979): 74–79.

28. See, for instance, "It's the Number of Solutions That Is the Problem for Y2K Bug," *Alexander's Gas and Oil Connections*, October 27, 1998, http://www.gasandoil.com/news/features/b381355656d77cfe6af52a81d37bc813, accessed July 21, 2017.

29. Martin Wainwright, "NHS Faces Huge Damages Bill after Millennium Bug Error," *Guardian*, September 13, 2001, https://www.theguardian.com/uk/2001/sep/14/martinwainwright, accessed July 21, 2017.

30. Dennis Dutton, "It's Always the End of the World as We Know It," *New York Times*, December 31, 2009, A29.

SUGGESTED FURTHER READING

This is not a complete bibliography, and it does not contain archival sources or those mentioned in the notes. Rather, it is intended as an introductory, mostly English-language bibliography for the beginning student (though I have, by necessity, included some French and German sources).

General

The "clock question" has a long and rich literature. It was perhaps first stated by Karl Marx in an 1863 letter to Friedrich Engels, where he identified the clock as the first applied use of the machine, "from which the whole theory of production of regular motion was developed" (*The Letters of Karl Marx*, ed. and trans. Saul K. Padover [Englewood Cliffs, New Jersey: Prentice-Hall, 1979], 168). However, there are a few fundamental works on this topic that bear special mention. Lewis Mumford's *Technics and Civilization* (New York: Harcourt Brace, 1934), which sees mentalities and machines as shaping one another, was important for framing the question today and linking it to the history of science and technology: "The clock . . . is a piece of power machinery whose 'product' is seconds and minutes: by its essential nature it dissociated time from human events and helped created the belief in an independent world of mathematically measurable sequences: the special world of science" (15). Carlo Cipolla's *Clocks and Culture 1300–1700* (New York: W. W. Norton, 2003; original 1967) is also fundamental. David Landes's *Revolution in Time* (Cambridge, MA: Harvard University Press, 1983) details the development of timekeeping technology and its social and economic effects. Interestingly, Landes's son, Richard, with whom I studied at Boston University, researched millennialism—theological ideas about the end of time. On the latter subject, see especially Norman Cohn's *Pursuit of the Millennium* (New York: Oxford University Press, 1990). In the realm of philosophy Gerald James Whitrow's *What Is Time?* (Oxford: Oxford University Press, 1972) is a classic account of ideas of time, while his *Time in History* (Oxford: Oxford University Press, 1988) details the historical development of ideas about time.

In my own field, medieval studies, other critical (if dated) pieces on ideas of time that established the premodern world as a timekeeping "other" are Marc Bloch, *La société féodale*, 2 vols. (Paris: Albin, 1939–1940), 1.117; Lucien Febvre, "Temps flottant, temps dormant," in *Le problème de l'incroyance:*

La religion de Rabelais (Paris: Albin, 1942), 426–434; linking time measurement and medieval capitalism is Jacques Le Goff, "Merchant's Time and Church's Time," in *Time, Work, and Culture in the Middle Ages*, translated by Arthur Goldhammer (Chicago: University of Chicago Press, 1980), 29–42. For a counter to these, see Gerhard Dohrn-van Rossum's critique in his extremely important *The History of the Hour*, translated by Thomas Dunlap (Chicago: University of Chicago Press, 1996). Finally, for a good historiographical overview of the social use of time, see Peter Burke, "Reflections on the Cultural History of Time," *Viator* 35 (2004): 617–626.

General works on technology and society include Ian Goldin and Chris Kutarna's *Age of Discovery: Navigating the Risks and Rewards of Our New Renaissance* (New York: St, Martin's, 2016), which is a good overview of innovation and its social effects, and Jo Ellen Barnett's *Time's Pendulum: From Sundials to Atomic Clocks, the Fascinating History of Timekeeping and How Our Discoveries Changed the World* (New York: Harcourt Brace, 1999) is similarly a recent history of timekeeping technology.

Chapter 1. Scholars and Spheres

Many of the ideas in this chapter were derived from conversations I had during the first peer-review cycle of this book with Professor Bert Hall, emeritus at the University of Toronto, and from Professor Hall's unpublished notes. On the actual scientific history, James Evans's *The History and Practice of Ancient Astronomy* (Oxford: Oxford University Press, 1998) was invaluable. More specifically, on the Antikythera Mechanism, see Derek J. de Solla Price, "An Ancient Greek Computer," *Scientific American* 200, no. 6 (June 1959): 60–67, and "Gears from the Greeks: The Antikythera Mechanism: A Calendar Computer from ca. 80 B.C.," *Transactions of the American Philosophical Society* 64, no. 7 (1974): 1–70; as well as Phillip Ball, "Complex Clock Combines Calendars," *Nature* 454, no. 7204 (July 30, 2008): 614–617. The best recent work on the Antikythera Mechanism is Alexander Jones, *A Portable Cosmos: Revealing the Antikythera Mechanism, Scientific Wonder of the Ancient World* (New York: Oxford University Press, 2017).

Trigonometry was long thought to be a Greek invention; on Mesopotamian origins, see Daniel F. Mansfield and N. J. Wildberger, "Plimpton 322 Is Babylonian Exact Sexagesimal Trigonometry," *Historia Mathematica Historia Mathematica* 44, no. 4 (November 2017): 395–419; uncorrected proof available at https://doi.org/10.1016/j.hm.2017.08.001, accessed August 24, 2017. On the Ach Valley engraving, see Michael Rappenglück, "The Anthropoid in the Sky: Does a 32,000 Years Old Ivory Plate Show the Constellation Orion Combined with a Pregnancy Calendar?," in *Symbols, Calendars and Orientations: Legacies of Astronomy in Culture; IXth Annual Meeting of the European Society for Astronomy in Culture (SEAC)*, edited by Mary Blomberg, Peter E. Blomberg, and Göran Henriksson (Uppsala: Uppsala University, 2003), 51–55. On other ancient observatories, see A. César González-Garcia, "Carahunge: A Critical Assessment," in *Handbook of Archaeoastronomy and Ethnoastronomy*, edited by Clive L. N. Ruggles (New York: Springer Science+Business Media, 2004), 1453–1460; and Ivan Ghezzi and Clive Ruggles, "Chankillo: A 2300-Year-Old Solar Observatory in Coastal Peru," *Science* 315, no. 5816 (2007): 1239–1243.

My chief source on Egyptian astronomy was Miguel Ángel Molinero Polo, "A Bright Night Sky over Karakhamun: The Astronomical Ceiling of the Main Burial Chamber in TT 223," in *Tombs of the South Asasif Necropolis: Thebes, Karakhamun (TT 223), and Karabasken (TT 391) in the Twenty-fifth Dynasty*, edited by Pischikova Elena (Cairo: American University in Cairo Press, 2014), 201–238. On Jewish astronomy, see Ari Belenkiy, "Astronomy of Maimonides and Its Arabic Sources," in *Cosmology across Cultures ASP Conference Series*, vol. 409, *Proceedings of the Conference Held 8–12 September 2008*, edited by José Alberto Rubiño-Martín, Juan Antonio Belmonte, Francisco Prada, and Antxon Alberdi (San Francisco: Astronomical Society of the Pacific, 2009), 188–202. On ancient Indian astronomy, see Walter Ashlin Fairservis, *The Harappan Civilization and Its Writing: A Model for the Decipherment of the Indus Script* (Leiden: E. J. Brill, 1992), 60–66; V. N. Tripathi, "Astrology in India," in *Encyclopaedia of the History of Science, Technology, and Medicine in Non-Western Cultures*, 2nd ed., edited by Helaine Selin (New York: Springer, 2008), 264–267; and S. Balachandra Rao, *Indian Astronomy: An Introduction* (Hyderabad: Universities Press, 2000). On Mesoamerican calendars, see Prudence M. Rice, *Maya Calendar Origins: Monuments, Mythistory, and the Materialization of Time* (Austin: University of Texas Press, 2007).

The fundamental author on Chinese timekeeping is Joseph Needham, especially *Science and Civilization in China* (Cambridge: Cambridge University Press, 1954), and *Heavenly Clockwork: The Great Astronomical Clocks of Medieval China* (Cambridge: Cambridge University Press, 1986). See also his *The Grand Titration: Science and Society in East and West* (Toronto: University of Toronto Press, 1969), 16. I am also indebted to Landes's critique thereof. On Chinese astronomy and the hours, see Mitsuru Sôma, Kin-aki Kawabata, and Tanikawa Kiyotaka, "Units of Time in Ancient China and Japan," *Publications of the Astronomical Society of Japan* 56, no. 5 (October 25, 2004): 887–904.

Chapter 2. Cities and Clocks

Probably the best recent work is Dohrn-van Rossum's aforementioned *History of the Hour*. See also my chapter with Denis Casey, "Time and Timekeeping," in *Handbook of Medieval Culture* edited by Albrecht Classen (Berlin: De Gruyter, 2015) 3: 1657–1679. Older but still seminal is Lynn White Jr.'s *Medieval Technology and Social Change* (Oxford: Oxford University Press, 1962).

Other works used: Arthur Robert Green's still-useful *Sundials, Incised Dials or Mass-Clocks* (London: SPCK and Macmillan, 1926); Percival Price, *Bells and Man* (Oxford: Oxford University Press, 1983); Allan A. Mills, "The Mercury

Clock of the Libros del Saber," *Annals of Science* 45, no. 4 (July, 1988): 329–344, and "Seasonal-Hour Sundials on Vertical and Horizontal Planes, with an Explanation of the Scratch Dial," *Annals of Science* 50, no. 1 (January 1993): 83–93; and Lynn Thorndike Jr.'s "Invention of the Mechanical Clock about 1271 AD," *Speculum* 16 (1941): 242–243, and *The Sphere of Sacrobosco and Its Commentators* (Chicago: University of Chicago Press, 1949).

On medieval philosophy of time, see Dirk-Jan Dekker's various articles. On medieval science in general, Charles Homer Haskins's *Studies in Medieval Science* (Cambridge, MA: Harvard University Press, 1924) is foundational to the field. Edward Grant and John Murdoch, eds., *Mathematics and Its Applications to Science and Natural Philosophy in the Middle Ages: Essays in Honor of Marshall Clagett* (New York: Cambridge University Press, 1987) contains many useful essays. On the *Liber Nemroth*, see Haskins, "Nimrod the Astrologer," in *Studies in Medieval Science*; and David Juste, "On the Date of the Liber Nemroth," *Journal of the Warburg and Courtauld Institutes* 67 (2004): 255–257. European astronomical instruments and theory have been studied in detail by Emmanuel Poulle in his several works, especially his *Les instruments astronomiques du Moyen Age* (Paris: Société internationale de l'astrolabe, 1983), *Les instruments de la théorie des planètes selon Ptolémée: Équatoires et horlogerie planétaire du XIIIᵉ au XVIᵉ siècle*, 2 vols. (Geneva: Droz, 1980–1981), *Les sources astronomiques (textes, tables, instruments)* (Turnhout, Belgium: Brepols, 1984), *Les tables alphonsines, avec les canons de Jean de Saxe* (Paris: Editions du Centre national de la recherche scientifique, 1987–1988), and his more recent translation of Giovanni Dondi dell'Orologio's *Tractatus astrarii* (Geneva: Droz, 2003). John D. North has written several works on Wallingford, of which the best and most comprehensive is *God's Clockmaker* (New York: Bloomsbury, 2005); see also North's translations of Wallingford's treatises, *Richard of Wallingford* (Oxford: Oxford University Press, 1976). On notaries, see Steven A. Epstein, "Business Cycles and the Sense of Time in Medieval Genoa," *Business History Review* 62 (1988): 238–260. For two excellent works on changing time regimes in England, see Chris Humphrey, "Time and Urban Culture in Late Medieval England," in *Time in the Medieval World*, edited by Chris Humphrey, and W. Mark Ormrod (Woodbridge: Boydell and Brewer, 2001), 105–118; and Karen Smyth, "Changing Times in the Cultural Discourse of Late Medieval England," *Viator* 35 (2004): 435–454.

On intellectual foundations, see Pierre Duhem, *Le système du monde* (Paris, 1956), especially 7:439–441, E. A. Burtt's *The Metaphysical Foundations of Modern Science* (Garden City, NY: Doubleday, 1951); Ursula Coope, *Time for Aristotle* (Oxford: Oxford University Press, 2005); Helen S. Lang, *Aristotle's Physics and Its Medieval Varieties* (Albany: State University of New York Press, 1992); R. Sorabji, *Time, Creation and the Continuum: Theories in Antiquity and the Early Middle Ages* (Ithaca, NY: Cornell University Press, 1983).

Fundamental works on Islamic timekeeping are David King's marvelous *In Synchrony with the Heavens: Studies in Astronomical Timekeeping and Instrumentation in Medieval Islamic Civilization*, 2 vols. (Leiden: Brill 2004–2005); and George Saliba's *A History of Arabic Astronomy: Planetary Theories During the Golden Age of Islam* (New York: New York University Press, 1995). Donald Hill, trans., *The Book of Ingenious Devices* (Boston: Kluwer Academic, 1979), and *Arabic Water Clocks* (Aleppo: University of Aleppo, 1981) are interesting primary sources. For two good, brief works on transmission to the West, see my mentor, Thomas Glick's *Islamic and Christian Spain in the Early Middle Ages*, 2nd rev. ed. (Leiden, Brill, 2005), chapter 8, and his chapter, "Science in Medieval Spain: The Jewish Contribution in the Context of Convivencia," in *Convivencia: Jews, Christians, and Muslims in Medieval Spain*, edited by Vivian B. Mann, Thomas F. Glick, and Jerrilynn Denise Dodds (New York: George Braziller, 1992, 276–283). On the Damascus clock, see Finbarr Barry Flood, *The Great Mosque of Damascus: Studies on the Makings of an Umayyad Visual Culture* (Boston: Brill, 2000).

For a thorough historiographical overview of the "Needham question" as it stood in the mid-1990s, see Hendrik Floris Cohen, *The Scientific Revolution: A Historiographical Inquiry* (Chicago: University of Chicago Press, 1994), 378–505. I also drew on David S. Landes's rebuttal of Needham in *Revolution in Time*, and Geoffrey Ernest Richard Lloyd's treatment of Chinese historiography in *Ambitions of Curiosity* (Cambridge: Cambridge University Press, 2002). For another important perspective, see Nathan Sivin, "Why the Scientific Revolution Did Not Take Place in China—or Didn't It?," *Chinese Science* 5 (1982): 45–66; revised August 24, 2005 and found at http://web.nchu.edu.tw/pweb/users/hbhsu/lesson/8252.pdf, accessed July 5, 2017. Another take on this is Justin Lin, "The Needham Puzzle: Why the Industrial Revolution Did Not Originate in China," in *Economic Development and Cultural Change* 43, no. 2 (January 1995): 269–292; see also Etienne Balazs, *La bureaucratie celeste: Recherches sur l'économie et la société de la Chine traditionnelle* (Paris: Gallimard, 1988). Balazs blames centralized, stifling state control as choking innovation. I have personally been influenced by both David Landes and Jared Diamond's arguments in *Guns, Germs, and Steel* (New York: W. W. Norton, 1997): China's geography made early and continual unification practical, and periods of disorder were comparatively few; complimentary mentalities led to China being, in Needham's words, "homeostatic" but not "stagnant." Of course, I am not a Sinologist (and neither was Landes), and this is a book on timekeeping technology, not the comparative historiography of science, and so I must beg the reader's indulgence.

Chapter 3. Savants and Springs

Galileo has been translated many times; I used *Dialogues Concerning Two New Sciences*, trans. Henry Crew and Alfonso de Salvio (Norwich, NY: William Andrew, 1914). On his biography, Stillman Drake's work, such as his *Discoveries and Opinions of Galileo* (Garden City, NY: Doubleday, 1957), and *Galileo at Work: His Scientific Biography* (Chicago: University of Chicago Press, 1978) are still important. On the pendulum experiment, see Thomas B. Settle, "An Experiment in the History of Science" *Science* 133, no. 3445 (1961): 19–23.

For a thorough discussion of the Scientific Revolution, I recommend Steven Shapin, *The Scientific Revolution* (Chicago: University of Chicago Press, 1998). Edgar Zilsel expressed his famous thesis in *The Social Origins of Modern Science*, edited by Diederick Raven, Wolfgang Krohn, and Robert S. Cohen (Dordrecht: Kluwer Academic, 2000); it was recently reinvigorated by Pamela O. Long in her *Artisan/Practitioners and the Rise of the New Sciences, 1400–1600* (Corvalis: University of Oregon Press, 2011). On Riccoli and others, see Domenico Bertoloni Meli, *Thinking with Objects: The Transformation of Mechanics in the Seventeenth Century* (Baltimore: Johns Hopkins University Press, 2006), 131–134. On Santorio, see Arturo Castiglione, "The Life and Work of Santorio Santorio (1561–1636)," translated by Emilie Recht, *Medical Life* 38 (1931): 729–785. There are many good sources on the Gregorian calendar; see especially G. V. Coyne, Michael A. Hoskin, and Olaf Pedersen, eds., *Gregorian Reform of the Calendar: Proceedings of the Vatican Conference to Commemorate Its 400th Anniversary, 1582–1982* (Vatican City: Pontifical Academy of Sciences, 1983); and Bonnie J. Blackburn and Leofranc Holford-Strevens, *The Oxford Companion to the Year: An Exploration of Calendar Customs and Time-Reckoning* (Oxford: Oxford University Press, 2003). A good recent work on Huygens is physicist C. D. Andriesse's *Huygens: The Man Behind the Principle*, translated by Sally Miedema (Cambridge: Cambridge University Press, 2005). See also Hans van den Ende, *Huygens' Legacy: The Golden Age of the Pendulum Clock* (Castletown: Fromanteel, 2004). A fun little article on anchor clock escapements is Michael Headrick, "Origin and Evolution of the Anchor Clock Escapement," *Control Systems Magazine* 22, no. 2 (2002): 41–52. Lynn White Jr. covers spring-driven escapements in *Medieval Technology and Social Change*, 126–127, and Dohrn-van Rossum in *History of the Hour*, 121. On Jost Burgi, see his entry in Lance Day and Ian McNeil, eds., *Biographical Dictionary of the History of Technology* (New York: Routledge, 1996), 204–205. On Hooke's lost minutes, see Robyn Adams and Lisa Jardine, "The Return of the Hooke Folio," *Notes and Records of the Royal Society of London* 60, no. 3 (September 22, 2006): 235–239. On Floyer, see Jacob Rosenbloom, "The History of Pulse Timing with Some Remarks on Sir John Floyer and His Physician's Pulse Watch," in *Annals of Medical History* (New York: Paul B. Hoebler, 1922), 4:98. On the development of "absolute time," see Milič Čapek, "The Conflict between the Absolutist and the Relational Theory of Time before Newton," *Journal of the History of Ideas* 48 (1987): 595–608. For a background of the political milieu and how Newtonianism came to be associated with the party of the philosophes, see J. B. Shank's *The Newton Wars* (Chicago: University of Chicago Press, 2008).

Chapter 4. Navigators and Regulators

Much of the material in this chapter is found in the above fundamental works. The best-known modern treatment of the longitude problem, if not a scholarly one, is Dava Sobel's *Longitude: The True Story of a Lone Genius Who Solved the Greatest Scientific Problem of His Time* (New York: Walker, 1995). Landes (1983) also discusses Harrison. My task here is, of course, to retell this well-known story in an accessible and entertaining manner for those who might be hearing it for the first time.

On analemmic sundials, see Frederick W. Sawyer III, "Of Analemmas, Mean Time and the Analemmatic Sundial," *Bulletin of the British Sundial Society* 94, no. 2 (June 1994), 2–6, and 95, no. 1 (February 1995): 39–44. The most recent biography of Halley is Alan H. Cook, *Edmond Halley: Charting the Heavens and the Seas* (Oxford: Clarendon Press, 1998). See also David W. Hughes, "Edmond Halley, Scientist," *Journal of the British Astronomical Association* 95, no. 5 (August 1985): 193–204, on his relationship with Flamsteed. On the much-vilified Maskelyne, see Derek Howse, *Nevil Maskelyne: The Seaman's Astronomer* (Cambridge: Cambridge University Press, 1989), and also his *Greenwich Time and the Longitude*, rev. ed. (London: Philip Wilson, 2003). The website of the Royal Observatory and National Maritime Museum in Greenwich at http://www.rmg.co.uk/national-maritime-museum is also a fascinating resource. The Huygens BMP-2 is described in the auction catalog of Dr. Crott Auctioneers, http://www.uhren-muser.com/cn/documents/News_7-2013_US.pdf, accessed August 24, 2017. Jean Richer is little attested, but J. J. O'Connor and E. F. Robertson of the University of St. Andrews have his biography at http://www-history.mcs.st-andrews.ac.uk/Biographies/Richer.html, accessed August 24, 2017. On Matthias Wasmuth, see Wolfgang Köberer, "On the First Use of the Term 'Chronometer,' " *Mariner's Mirror* 102, no. 2 (2016): 203–206.

On decimal time, see Hector Vera, "Decimal Time: Misadventures of a Revolutionary Idea," *KronoScope* 9, nos. 1–2 (2009): 29–48. On industrialism, E. P. Thompson's "Time, Work-Discipline, and Industrial Capitalism," *Past and Present* 38 (1967): 56–97, is foundational and a useful continuation of Marx's ideas; Mumford, *Technics and Civilization*, and White, *Medieval Technology and Social Change*, are also central. William Radcliffe's *Origin of the New System of Manufacture, Commonly Called Power Loom Weaving* (London, 1828) was reprinted in J. F. C. Harrison, *Society and Politics in England, 1780–1960* (New York: Harper & Row, 1965). On medieval Genoa, see Steven A. Epstein, "Business Cycles and the Sense of Time in Medieval Genoa," *Business History Review* 62 (1988): 238–260. Many books have been written on Lowell, but see especially Thomas Dublin, *Women at Work: The Transformation of Work and Community in Lowell, Massachusetts 1826–1860* (New York: Columbia University Press, 1979). For a basic biography of Taylor, see Frank Barkley Copley, *Frederick W. Taylor, Father of Scientific Management* (New York: Harper and Brothers, 1923). His works were published together as

Principles of Scientific Management (New York: Harper & Brothers, 1913). There are a number of works on the Gilbreths and especially Lillian Gilbreth as a female academic pioneer, but see especially her autobiography, *As I Remember: An Autobiography* (Norcross, GA: Management and Engineering Press, 1998); and Frank Bunker Gilbreth Jr. and Ernestine Gilbreth Carey, *Cheaper by the Dozen* (New York: Thomas Y. Crowell, 1948, later made into a film), and *Belles on Their Toes* (New York: Thomas Y. Crowell, 1950).

The Duchamp quote comes from Katherine Kuh, ed., *The Artist's Voice: Talks with Seventeen Modern Artists* (Harper & Row, New York 1962), 81–93. For a good and readable brief history of Liebig's beef extract, see Clay Cansler, "Where's the Beef?," *Distillations*, Fall 2013/Winter 2014, https://www .chemheritage.org/distillations/article/where%E2%80%99s-beef, accessed August 24, 2017.

The history of electric clocks has mostly been written by amateur enthusiasts. Michel Viredaz' clock history site at http://www.chronometrophilia.ch/Electric -clocks/english.htm, accessed August 24, 2017, was very useful in preparing this section, as was the ClockDoc website at http://wp.clockdoc.org/, accessed August 24, 2017. Another great website with animations that helped me immensely was J. E. Bosschieter's "History of the Evolution of Electric Clocks," http://www.electric-clocks.nl/clocks/en/index.htm, accessed August 24, 2017. On Zamboni, see Massimo Tinazzi, "Perpetual Locomotive of Guiseppe Zamboni," http://www.brera.unimi.it/sisfa/atti/1996/tinazzi.html, accessed August 24, 2017; the page is quite scholarly and has an extensive bibliography. On Wheatstone and Bain, see Brian Bowers, *Sir Charles Wheatstone FRS: 1802–1875* (London: Science Museum, 2001), 154–155. On Jones's system, see David Glasgow, *Watch and Clock Making* (London: Cassell, 1885), 314–315. See also Charles Piazzi Smyth, *Astronomical Observations Made at the Royal Observatory, Edinburgh* (Edinburgh: Neill, 1871), 13: xxxi–xxxii; F. Hope Jones, *Electrical Timekeeping* (London: NAG Press, 1940), https://clockdoc.org/gs/handler /getmedia.ashx?moid=23814&dt=3&g=1, accessed August 24, 2017; J. Kieve, *The Electric Telegraph: A Social and Economic History* (Newton Abbot, Devon: David & Charles, 1973). Bain himself wrote *A Short History of the Electric Clocks* (London: Chapman and Hall, 1852; repr. London: Turner & Devereux, 1973).

On time zones, see Alvin Powell, "America's First Time Zone," *Harvard Gazette*, November 10, 2011, http://news.harvard.edu/gazette/story/2011/11 /americas-first-time-zone/, accessed August 24, 2017. On Dowd, see R. Newton Mayall, "The Inventor of Standard Time," *Popular Astronomy* 50 (April 1942): 204–208. William F. Allen's story has been told many times; his papers are housed at the New York Public Library, http://archives.nypl.org/mss/53, accessed August 24, 2017. See also Sandford Fleming's *Universal or Cosmic Time by Sandford Fleming, C.E., C.M.G., etc.: Together with Other Papers, Communications and Reports in the Possession of the Canadian Institute respecting the Movement for Reforming the Time-System of the World and Establishing a Prime Meridian as a Zero Common to All Nations* (Toronto: Council of the Canadian Institute, 1885). The Smithsonian also has a good page at http://americanhistory.si.edu/ontime/synchronizing/zones.html, accessed August 24, 2017. NOAA's website at https://oceanservice.noaa.gov/facts /international-date-line.html, accessed August 24, 2017, is very useful on the international date line. The Samoan adjustment of 2011 and the problems it caused were widely reported in the news at the time. See George Hudson's original proposal, "On Seasonal Time-Adjustment in Countries South of Lat. 30°," in James Hector, *Transactions and Proceedings of the New Zealand Institute* (Wellington: John McKay Government Printing Office, 1895), 28: 734.

Chapter 5. Rationalization and Relativity

There are a number of accessible explanations of Einstein for general readers, but the best is probably Einstein's own *Relativity: The Special and the General Theory* (Princeton, NJ: Princeton University Press, 2015). Robert Geroch's *General Relativity from A to B* (Chicago: University of Chicago Press, 1981) is also good, while my personal favorite is the illustrated *Introducing Relativity* written by Bruce Bassett and illustrated by Ralph Edney (London: Totem Books, 2002). I am not the first historian to make this comparison between Einstein and "modern" culture. See, for instance, Stephen Kern's comparisons between Einstein and Durkheim in his *The Culture of Time and Space, 1880–1918*, 2nd ed. (1983; Cambridge, MA: Harvard University Press, 2003), 19–20. For varied perspectives and bibliography on this, see Peter L. Galison, Gerald Holton, Silvan S. Schweber, eds., *Einstein for the 21st Century: His Legacy in Science, Art, and Modern Culture* (Princeton, NJ: Princeton University Press, 2008); and Gerald Holton and Yehuda Elkana, eds., *Albert Einstein, Historical and Cultural Perspectives: The Centennial Symposium in Jerusalem* (Princeton, NJ: Princeton University Press, 1984). Peter Galison's *Einstein's Clocks and Poincare's Maps: Empires of Time* (New York: W. W. Norton, 2003) is a good overview (though I don't discuss Poincare). Arthur Eddington published his observations as *Report on the Relativity Theory of Gravitation* (London: Fleetway Press, 1918); his biography is by Allie Vincent Douglas, *The Life of Arthur Stanley Eddington* (London: T. Nelson, 1956). Bertrand Russell's *The ABCs of Relativity* (another great explanation of Einstein) was published by Allen & Unwin in New York in 1958. Poul Anderson's *Tau Zero* was published in New York by Doubleday in 1970.

On Beeckman, see Klaas van Berkel, *Isaac Beeckman on Matter and Motion: Mechanical Philosophy in the Making* (Baltimore: Johns Hopkins University Press, 2013). On Rømer, see Laurence Bobis and James Lequeux, "Cassini, Rømer and the Velocity of Light," *Journal of Astronomical History and Heritage* 11, no. 2 (2008): 97–105. The seminal paper is I. Bernard Cohen, "Roemer and the First Determination of the Velocity of Light (1676)," *Isis* 31, no. 2 (1940): 327–379. On James Bradley, see Mari E. W. Williams, "Bradley, James (bap. 1692, d. 1762)," in *Oxford Dictionary of National Biography*, edited by H. C. G. Matthew and Brian Harrison (Oxford: Oxford University Press, 2004). Hippolyte Fizeau published "Sur les hypothèses relatives à l'éther lumineux" in *Comptes rendus hebdomadaires des séances de l'Académie des sciences* (Paris: Académie des sciences, 1851), 33: 349–355. On Michelson, see Dorothy Michael Livingston, *The Master of Light: A Biography of Albert A. Michelson* (Chicago: University of Chicago Press, 1973). Louis Essen published, with Keith Davy Froome, *The Velocity of Light and Radio Waves* (New York: Academic Press, 1969). For biographical information on Essen, see Alan Cook, "Louis Essen, O. B. E. 6 September 1908–24 August 1997," *Biographical Memoirs of Fellows of the Royal Society* 44 (1997): 143. On the atomic clock, I found Andrea Sella, "Essen's Clock," *Chemistry World*, January 27, 2014, https://www.chemistryworld.com/opinion/essens-clock/7017.article, accessed July 21, 2017. The original paper was Louis Essen and J. V. L. Parry, "An Atomic Standard of Frequency and Time Interval: A Cæsium Resonator," *Nature* 176, no. 4476 (1955): 280. On the 1972 NIST experiment, see Ken Evenson et al., "Speed of Light from Direct Frequency and Wavelength Measurements of the Methane-Stabilized Laser," *Physical Review Letters* 29, no. 19 (November 6, 1972): 1346. On NIST radio stations, see Glenn K. Nelson, Michael A. Lombardi, and Dean T. Okayama, "NIST Time and Frequency Radio Stations: WWV, WWVH, and WWVB" (2005). http://tf.nist.gov/timefreq/general/pdf/1969.pdf, accessed July 21, 2017.

On Kelvin and the age of the earth, see the aptly titled Joe Burchfield, *Lord Kelvin and the Age of the Earth* (Chicago: University of Chicago Press, 1975), 94 and 140. On Hubble, see Gale Christianson, *Edwin Hubble: Mariner of the Nebulae* (New York: Farrar, Straus and Giroux, 1995). On Hipp, see Helmut Kahlert, "Lorenz Bob und Matthäus Hipp," in *Alte Uhren und moderne Zeitmessung* (Munich: Callwey Verlag, 1987), 22:4; and Hans Rudolf Schmid "Hipp, Matthäus," in *Neue Deutsche Biographie* (Berlin: Duncker & Humblot, 1972), 9:199–200. Alfred Niaudet-Bréguet published "Application du diapason à l'horlogerie" in *Comptes rendus de l'Académie des Sciences* (Paris, 1866), 63:991–92; see also Françoise Birck and André Grelon, eds., *Un siècle de formation des ingénieurs électriciens: Ancrage local et dynamique européenne; l'exemple de Nancy* (Paris: Éditions de la Maison des sciences de l'homme, 2006), 407.

On quartz timepieces, my information on Warren was *Henry Ellis Warren: A Biographical Memoir* (New York: American Historical Company), http://www.telechron.com/telechron/warren_bio.pdf, accessed July 20, 2017. On Cady, see Warren P. Mason, "Professor Walter G. Cady's contributions to piezoelectricity and what followed from them," *Journal of the Acoustical Society of America* 58, no. 2 (August 1975): 301–309); and Shaul Katzir, "From Ultrasonic to Frequency Standards: Walter Cady's Discovery of the Sharp Resonance of Crystals," *Archive for History of Exact Sciences* 62, no. 5 (September 2008): 469–487. The original research paper was Warren Marrison and J. W. Horton, "Precision Determination of Frequency," *Proceedings of the Institute of Radio Engineers* 16, no. 2 (February 1928), 137–154; see also Warren Marrison, "The Evolution of the Quartz Crystal Clock," *Bell System Technical Journal* 27, no. 3 (July 1948): 510–588. On the development of wristwatches, see Carlene Stephens and Maggie Dennis, "Engineering Time: Inventing the Electronic Wristwatch," *British Journal for the History of Science* 33, no. 4 (2000): 477–497. On the development of watches in general, see the epilogue to Alexis McCrossen, *Marking Modern Times: A History of Clocks, Watches, and Other Timekeepers in American Life* (Chicago: University of Chicago Press, 2016).

On the ephemeris year, Willem de Sitter's original publication was "On the Secular Accelerations and the Fluctuations of the Longitudes of the Moon, the Sun, Mercury and Venus," *Bulletin of the Astronomical Institutes of the Netherlands* 4 (1927): 21–38. On Y2K, see Ulrich Windl, David Dalton, Marc Martinec, and Dale R. Worley, "The NTP FAQ and HOWTO: Understanding and Using the Network Time Protocol (a First Try on a Non-technical Mini-HOWTO and FAQ on NTP)," https://www.eecis.udel.edu/~ntp/ntpfaq/NTP-a-faq.htm, accessed July 21, 2017. See also Robert W. Bemer, "Time and the Computer," *Interface Age Magazine* 4, no. 2 (February 1979): 74–79). Another forecast, closer to the date: "It's the Number of Solutions That Is the Problem for Y2K Bug," *Alexander's Gas and Oil Connections*, October 27, 1998, http://www.gasandoil.com/news/features/b381355656d77cfe6af52a81d37bc813, accessed July 21, 2017. On the medical mistakes in the UK, see Martin Wainwright, "NHS Faces Huge Damages Bill after Millennium Bug Error," *Guardian*, September 13, 2001, https://www.theguardian.com/uk/2001/sep/14/martinwainwright, accessed July 21, 2017; and, for some perspective, see Dennis Dutton, "It's Always the End of the World as We Know It," *New York Times*, December 31, 2009, A29.

Abelard, Peter, 28
absolute and relative time, 69–71,
   77–78, 105–109, 112
accuracy, 7, 8, 51, 163–172
Accutron watch, 164, 166
AD. *See* anno Domini
Adams, Walter Sydney, 157
Alfonso X "the Wise" of Castile and
   Leon, 55
Algamest, 15–16, 36, 44, 55. *See also*
   Ptolemy
Allen, William Frederick, 140–142
AM, 32
analog computer, 2, 102
analema, 118, 180–184
anchor escapement, 91
Anderson, Poul, 152–153
Anglicus, Robertus, 54–55
anno Domini, 10
Antikythera Mechanism, 1–3, 7
antisynchronization, 88, 90
apparent solar time, 117
Apollo program. *See* space travel
Arabic timekeeping, 5, 44–46; intercala-
   tion forbidden, 27, 51; Islamic prayer
   times in, 44, 51–52, 138; water clocks
   in, 49–52
Arkwright, Richard, 131
Aristotle, 66–71, 82, 84–85, 108–109
Arnold, John, 126
Asia, timekeeping systems in, 24–25, 34.
   See also China, time; Japan
astrolabes, 44–46, 185–193
astronomical timekeeping, 9, 11, 12–23,
   35–36, 42–43, 167–168; Chinese, 19,
   34; Islamic, 44. *See also* astrolabes
atomic clocks, 4, 158, 168–172
Augustine of Hippo (St. Augustine), 6,
   31, 78

Augustus (Roman emperor), 29
automata, 100–102

Babylonian timekeeping, 9, 13–14
base-12. *See* duodecimal timekeeping
Bacon, Roger, 55–56
Bain, Alexander, 140
balance spring, 95, 95–98
balance wheel, 95, 95–98
Barrow, Isaac, 106–107
BC. *See* anno Domini
Bede, 10, 160
Bemer, Robert, 174
Beirut, 138
Bellair, Charles, 103
bells, 39–41
Benedictine Rule, 31, 33
Bernardino, Telesio, 106
Bliss, Nathaniel, 125
Book of Astronomy (*Liber Nemroth*),
   42–43
Books of Astronomical Knowledge
   (*Libros del saber de astronomia*), 55
Bracciolini, Poggio, 105–106
Bradley, James, 122, 124–125, 155
Buridan, Jean, 68–71, 131

Cady, Walter Guyton, 164
Caesar, Julius, 28–29. *See also* Gregorian
   calendar reform
calendars, 11; ancient Egyptian, 12–13;
   ancient Greek, 14–15; ancient
   Mesopotamian, 13–14; ancient
   Roman, 23–24; Carolingian, 31; in
   different cultures, 24–30; Jewish,
   25–26
candles, 37, 46, 67
Čapek, Milič, 106
capitalism, 128–137

Carolingian Empire, 31, 42
Cartwright, Edmund, 131
Cassian, John (saint), 35
Cassini, Giovanni Domenico, 154
cell phones. *See* mobile phones
Chanquillo, 12
Charles V, King of France, 58, 63, 64
Charles the Great (Charlemagne), 31,
    49, 51
Charleton, Walter, 106
China: calendars in, 24–25; Jesuit
    missionaries in, 104–107; hours in,
    34; possible invention of the mechani-
    cal clock in, 56–60; timekeeping in, 9
Christina, Grand Duchess of Tuscany, 79
Christine de Pisan, 46–47
Christianity: bells, 39–41; calendar,
    29–31; Feria days, 25–26; liturgical
    hours in, 32–34, 37–38, 39–41; moral
    importance of timekeeping in, 37,
    128. *See also* Benedictine Rule;
    Gregorian Calendar Reform; Jesuits;
    Millennialism; Protestantism
chronometer, 89, 112, 116–117,
    122–128
Clavius, Christopher, 84–88
Cleomedes, 19
Clepsydra. *See* water clock
clockmakers and watchmakers.
    See Arnold, John; Coster, Salamon;
    Earnshaw, Thomas; Graham, George;
    Heinlein, Peter; Hetzel, Max; Jaquet-
    Droz, Pierre; Jeffries, John; Mudge,
    Thomas; Ritchie, James; Tompion,
    Thomas; Vaucanson, Jacques de;
    Watson, Samuel; Women in horology
clocks, 3, 6, 42; adoption of the
    mechanical clock in France, 63;
    adoption of the mechanical clock in
    Italy, 61–62; evidence in scholastic
    philosophy, 65, 68–70; invention of
    mechanical clock, 54–65; possible
    invention of the mechanical clock in
    China, 56–60; regulation of urban life
    by, 64–65
Coleta (saint), 74
colonialism. *See* imperialism
computer time, 172–173. *See also* Y2K
    problem

Constantine (Roman Emperor), 30
Comstock, Daniel Frost, 152
constellations, 4, 11, 13, 17, 21. *See also*
    astronomical timekeeping
Copernicus, Nicolaus, 16, 75, 79,
    81, 83
Coster, Salamon, 91
Council of Nicea, 30
cross-beat escapement, 93
Curie, Marie, 164
Curie, Pierre, 164
cycloids, 91

Darwin, Charles, 161–162
day, 4, 9, 20. *See also* mean solar day.
deadbeat escapement, 91–92
decimal time, 127
Descartes, René, 109, 147
daylight savings, 141
digital displays, 166
Diocletian (Roman emperor), 27
Ditton, Humphrey
Dondi, Jacobo and Giovanni, 62
Dowd, Charles F., 140
Duchamp, Marcel, 136
Duodecimal timekeeping, 10
Durand, Guillaume, 38, 41, 63
duration, 46–48, 77–78
Dye, David, 168

Earnshaw, Thomas, 126–127, 128
ease of use, 7, 19
Earth, age of, 160–162
ecliptic, 21
Eddington, Arthur, 157
Egyptian timekeeping, 9, 12–13
Einstein, Albert, 147–153, 156–162
electric clock, 130, 139–140, 163–167
electrostatic clock, 139
Engels, Friedrich, 130
Enlightenment, 81, 83. *See also* Progress,
    idea of; *and names of individual
    thinkers*
entrainment, 90
ephemeris time, 167–168, 170
epoch, 167, 172–173
Epstein, Steven, 131
equal hours, 73. *See also* sidereal hour
Eratosthenes of Cyrene, 19

Essen, Louis, 155–156, 168–169
equation clocks, 104, 119
equation of time, 104, 116, 118
equinox, 20
escapement, 49, 85–88. *See also*
  deadbeat escapement; pendulum;
  Virge and Foliot

factories, 130–135
fencing, 67
Fizeau, Armand Hippolyte Louis, 155,
  163–164
Flamsteed, John, 119–122, 154
Fleming, Sandford, 142
Floyer, John, 99
Foucault, Léon, 155
Franklin, Benjamin, 130, 141
Frisius, Gemma, 115
Froissart, Jean 63
Fusee, 92–93

Galileo (Galileo Galilei), 4, 75–85, 86,
  92, 100, 114, 151, 155, 194–195
Galileo, Vincenzo, 80, 85
Gan De, 19
Gassendi, Pierre, 106
General Conference on Weights and
  Measures, 167–168, 170
Gerbert of Aurillac, 46
Gerlach, Walther, 168
Gilbreth, Frank, 135–137
Gilbreth, Lillian Moller, 135–137
Giebert, George Christian, 137
Global Positioning System (GPS), 158
Golein, Jean, 63
Goodman of Paris, 47
Graham, George, 92
grandfather clock. *See* longcase clock
Greece. *See* Hellenistic World, timekeep-
  ing in
Greenwich mean time, 120, 127–128,
  139, 142–146, 167
Gregorian Calendar Reform, 84–87, 174
Gregory of Tours, 35, 160

Halde, Jean-Baptiste de, 105–107
Halley, Edmond, 120, 122
Hargreaves, James, 131
harmonic oscillators, 88, 95, 164–165

Harrison, John, 112, 122–128, inven-
  tions, 121
Harrison, William, 123–125, 126
Heinlein, Peter, 93
Hellenistic World, timekeeping in, 1–3,
  14–20
Heloise, 38
Hertz, Heinrich, 164
Hesiod, 14
Hetzel, Max, 164, 166
Hooke, Robert, 4, 77, 95–98
Horologium, 42–43
horology, 17
Horton, J. W., 165
hour. *See* China, hours in; liturgical
  hour; sidereal hour
Hourglasses, 47
Hubble, Edwin, 162
Huygens, Christiaan, 77, 85, 95, 97,
  103, 114, 116, 117, 119

Imperialism, 5, 129, 138, 139, 143–144
India, timekeeping in, 9, 10; calendars
  in, 28; railway time in, 139
Industrial Revolution, 128–133
inertial frames, 150–151
intercalation, 25, 27, 84–85
International Astronomical Union, 167
international date line, 145
International Meridian Conference,
  142–143
Isidore of Seville, 37
isochronism, 80, 95
Islam. *See* Arabic timekeeping; Islamic
  calendar
Islamic calendar, 27–28
Italian hours, 62

Japan, timekeeping in, 143–144
Jaquet-Droz, Pierre, 101–102
Jeffries, John, 123
Jesuits, 80, 101, 104–107. *See also*
  Clavius, Christopher; Halde, Jean-
  Baptiste de; Ricci, Matteo; Riccioli,
  Giovanni Battista
Johannes de Sacrobosco, 54
Jones, R. L., 140
Judaism, calendars in, 25–27
Jupiter, moons of, 76, 108, 115

kalends, 25
Kay, John, 131
Kelvin, Lord (William Thomson), 161–162
Kepler, Johannes, 161

Landes, David, 129, 132
Liebig, Justus von, 137
Lilius, Aloysius, 84–88
liturgical hours. *See* Christianity
Locke, John, 81, 109
Lombe, John, 131
longcase clock, 92
longitude, 113–117; prizes for determining, 111–113. *See also* chronometer; lunar-distance method
Lorentz, Heinrich, 153
Lorentz factor, 153
Lowell, Francis Cabot, 132
Lucretius, 106
lunar-distance method, 115, 122–128
lunar timekeeping, 22
Lund, John Alexander, 140

Mach, Ernst, 151
mainspring, 94
Marconi, Guglielmo, 164
Marrison, Warren, 165
Marx, Karl, 130
Maskelyne, Nevil, 112, 120, 122–128, 144
master clock. *See* primary and secondary clocks
Mayan calendar. *See* Mesoamerica, calendars in
Mayer, Tobias, 122, 125
mean solar day, 16, 20–21, 117–120, 144–145
mean solar time. *See* mean solar day
mechanical clock. *See* clocks
Mercator, Geraldus, 115
Mesoamerica, calendars in, 28–30
Metonic calendar, 23
metric timekeeping, 10
Middle East. *See* Arabic calendar; Arabic timekeeping
Michaelson, Albert, 155
millennialism, 160–161, 175
"mill girls," 132–133

Mills, Alan, 43
mobile phones, 5, 167, 173
modernity. *See* Progress, idea of
month, 10, 11, 22; names of, 23
Moon. *See* lunar timekeeping; month
Morin, Jean-Baptiste, 115
Mudge, Thomas, 98
Muslim world. *See* Arabic timekeeping
Muybridge, Eadweard, 136

National Institute of Standards and Technology, 156, 170
Needham, Joseph, 56
Newton, Isaac, 77–78, 81, 105–109, 112–113, 116, 122, 147, 149, 154, 156, 161, 167
Niaudet-Bréguet, Alfred, 163
Nicholas of Cusa, 71, 77, 93, 101

Ogle, Vanessa, 138, 143
Olympiads, 2
Oresme, Nicole, 69–71, 113

Paris, 39–41
pendulum, 78–93, 194–195
Pepys, Samuel, 95
Phillip the Good, Duke of Burgundy, 93, 101
philosophy of time: medieval, 68–71; relativity, 159–162. *See also* absolute and relative time
Pierce, George Washington, 164–165
piezoelectric effect, 164
planets, 19–20, 22; Mercury and relativity, 156. *See also* Jupiter, moons of
Plato, 69–70, 78
pocket watches, 92–98, 129
PM, 32
prayer, Christian times for. *See* Christianity
precision, 7, 8, 51, 163–172
primary and secondary clocks, 139–140, 163
process time, 172
progress, idea of, 5, 129–130, 147, 159
Protestantism, 79–80
Ptolemaic model of the universe, 16, 21–22

Ptolemy, 15–17, 21, 36
pulsilogium, 82–85

quartz oscillators, 164–166, 168

Rabi, Isidor, 168
Radcliffe, William, 129
"radium girls," 134
railroads and railroad time, 130, 137–142
Ramis, Alois, 139
regulator clock. *See* primary and secondary clocks
relative time. *See* absolute and relative time
relativity, 147–161
relativity of simultaneity, 151
religion and timekeeping, 11. *See also* Arabic timekeeping, Christianity, Jesuits, Judaism, calendars in
remontoire, 93
Ricci, Matteo, 104–107
Riccioli, Giovanni Battista, 80–83, 84
Richard of Wallingford, 60–61
Richter, Jean, 116, 154
Ritchie, James, 140
Rome: calendars in, 23–29; dates in, 26–27; hours in, 32–33; week, 24
Rømer, Ole, 154
Ronalds, Francis, 139
Royal Society, 80–81
Rudd, Robert James, 140
Russell, Bertrand, 159–160

Santorio Santorio, 82–84, 92
Scott, David, 77
Scientific Revolution, 78–83, 84, 102, 160, 228
secondary clock. *See* primary and secondary clocks
seconds, 167–168
Shovell, Cloudesley, 110–111, 114
sidereal day, 16
sidereal hour, 32–36
sidereal month, 22–23
simplification, 2, 7, 103
Sitter, Willem de, 167
slave clock. *See* primary and secondary clocks

Smith, A. C. Gordon, 155
Smith, Adam 128
Sobel, Dava, 112
solar year. *See* tropical year
solstices, 12, 20
space travel, 77, 158, 166, 198
Spain, 144
speed of light, 151–156
Spencer, Herbert, 129
spring-driven clocks, 92–98
Standard time, 130, 137–146
stackfeed, 92
Steinheil, Carl August von, 139
Stern, Otto, 168
Stevin, Simon, 77
Stonehenge, 12
strob escapement, 60–61
Stroh, Augustus, 140
stopwatches, 98–100
Su Song, 58–60
Sun. *See* Day; tropical year
sundials, 32–33, 43, 103; Chinese, 34
Suso, Heinrich, 58, 63
Swatch, 166–167
Switzerland, 102
sympathetic magic, 115–116
synchronous clock, 163
synodic month, 22–23

Taylor, Frederick Winslow, 133–135
technology, 3
telegraphs, 137
theology, 10–11
Thompson, E. P., 130
Thomson, William. *See* Kelvin, Lord
time, nature of, 6–7
time dilation, 198. *See also* Lorentz factor; Relativity
timekeeping, 3; origin of systems, 9
time zones, 114, 142–146
timing devices, 48. *See also* stopwatch
Tolman's paradox, 198
Tomlinson, Ray, 5
Tompion, Thomas, 91
torsion spring. *See* mainspring
Towneley, Richard, 91
travelers' Dilemma. *See* Nicole Oresme
Treatise on the Sphere (*De Sphera*), 54
tropical year, 20

tropics, 20, 21
Truitt, Elly, 101
tuning-fork clock, **163**
"twins paradox," 151

Umiliana de' Cerchi, 53
unequal hours, 21, 32, 41–42
universe, age of, 162
universities, medieval, 65–66
urban clock. *See* clocks
Ussher, James, 160–161

Vaucanson, Jacques de, 101
Vespucci, Amerigo, 114
virge and foliot escapement, 59, 71–73
Vitruvius (Marcus Vitruvius Pollio), 33, 42, 49, 180–184

wall clock time, 172
water clocks, 49–54, 76–77. Arabic, 49–52; Chinese, 34
Watson, Samuel, 99
Watt, James, 131

Weber, Max, 128
Werner, Johannes, 115
Wheatstone, Charles, 139–140
Whiston, William, 116
Whitrow, G. J., 9
William of Ockham, 68–71
women, 134; as workers, 131–133; and wristwatches, 165. *See also* Coleta (saint); Christina, Grand Duchess of Tuscany; Christine de Pisan; Curie, Marie; Gilbreth, Lillian Moller, 135; *Goodman of Paris*; Heloise; "mill girls"; Umiliana de' Cerchi
work time, 64–65, 128–137
wristwatches, 165–167

year, 20–22
year 2000, 144
Y2K problem, 8, 149, 173–175

Zamboni, Giuseppe, 139
Zilsel, Edgar, 79. *See also* clockmakers
zodiac, 17
Zorats Karer, 12